Orchid
Muse

ALSO BY ERICA HANNICKEL

Empire of Vines: Wine Culture in America

Orchid
Muse

A History of Obsession in Fifteen Flowers

ERICA HANNICKEL

W. W. NORTON & COMPANY
Independent Publishers Since 1923

This book is intended as a general information resource. The author is not a professional orchid grower. Neither the general notes at the end of each chapter nor the "Top Fifteen Tips for Growing Orchids" at the end of the book are intended to substitute for professional advice about choosing and growing particular kinds of orchids.

For information about permission to reproduce selections from this book, write to Permissions, W. W. Norton & Company, Inc., 500 Fifth Avenue, New York, NY 10110

For information about special discounts for bulk purchases, please contact W. W. Norton Special Sales at specialsales@wwnorton.com or 800-233-4830

Manufacturing by TransContinental
Book design by Chris Welch Design
Production manager: Julia Druskin

ISBN 978-0-393-86728-2

W. W. Norton & Company, Inc., 500 Fifth Avenue, New York, N.Y. 10110
www.wwnorton.com

W. W. Norton & Company Ltd., 15 Carlisle Street, London W1D 3BS

1 2 3 4 5 6 7 8 9 0

FEBRUARY 2023

for Jason and Miles
in unending love, thanks,
and snuggles

Contents

✧

Part III. The Business of Beauty

Part IV. Orchid Culture

Appendices

Introduction

Orchidelirium

There is nothing quite like the raw sexuality of an orchid. We're reminded of their carnality by the flower's etymological root: the family is named for *orchis*, an ancient Greek word for testicle. The vanilla orchid (*vainilla* being "little pod" in Spanish, stemming from the Latin for vagina) references the impregnated bean that gives birth to vanilla seeds, the swollen pod's luscious scent long thought to be an aphrodisiac. Humans throughout time have imagined a spectrum of sexual organs in these blooms.

Orchids' lascivious architecture and otherworldly beauty have given rise to generations of orchid fanatics, and, at times, general orchidelirium. *Orchidomania* and *orchidelirium*—terms used since the nine-

Jacob Stead, *Orchids* (2019).

teenth century—capture the obsessive pursuits of orchid hunters and lay public alike who were driven to collect new orchids from dangerous yet environmentally fragile locations and devote their lives to their cultivation. One such moment of orchid fever occurred in the early Victorian era, sweeping through Europe in the 1850s and soon catching fire in America. Individual orchids sold for more than $1,000 ($35,000 today), orchid shows offered enormous prizes for choice specimens, and the flowers were embroidered on dresses, incorporated into elaborate jewelry designs, and painted on wallpaper. Stories emerged from every corner of the globe echoing the same cautionary note—that many a fortune had been made in orchids, but also many lives lost.

We might think that something inherently bewitching about orchids—something beyond description—lies behind this obsession. But it took a confluence of several cultural factors for orchidelirium to ignite. The industrial revolution created the fortunes of wealthy patrons who then employed orchid hunters to savage far-flung jungles for flowers. Improvements in steel and glass technology combined to create much larger and more elaborate greenhouses for pampering tropical orchids in cold climates. More efficient printing presses (producing books full of alluring color prints) and higher rates of literacy meant orchid knowledge was more widely accessible. Scientists became more skilled in classifying difficult flowers, and horticulturists perfected long-distance transportation for plants from their native environments and better growing techniques at home. And finally, Victorian women's fashion and home décor trended toward the ornate, shot through with romantic greenery.

It wasn't just the Victorians who suffered from orchidelirium. Orchids were revered in ancient China and feudal Japan and were the subject of lascivious poetry in Enlightenment Europe. On islands throughout the western Indian Ocean vanilla plantations boomed, far from the orchid's home in Mexico. Orchids even had a role in the nineteenth century antislavery effort, and their cultivation has been a solace for various groups seeking societal acceptance throughout time.

However magical orchids themselves may appear to you and me, I find the tales behind orchids equally as captivating. The following chapters recount the stories of more than twenty people who helped kindle orchidomania around the world, and those that keep ardor for orchids smoldering. For too long, these stories have been limited to a handful of explorers and scientists whose names many orchids bear. But orchids have a lot more to tell us beyond their names—they provide insight into human history, reveal intricate personal and international relationships, stand as dynamic political symbols, and reveal our constantly changing taste in beautiful living things. If we extend our gaze to the women and people of varied backgrounds who also contributed to orchid lore—sometimes behind the scenes, but often at center stage—we'll uncover the tenacity of orchid hunters, the lust of queens and empresses for perfect flowers, unexpected anthophiles from all walks of life, and artists' and authors' quests to capture the libidinous flowers in their work. In the struggle to preserve Earth's ecosystems, those charismatic, colorful orchids often stand in for millions of less-showy plants, signifying the sweeping beauty and complexity of nature lost to land conversion, poaching, and climate change. Spanning four centuries and more than a dozen countries, *Orchid Muse* uncovers the beautiful, hidden, and sometimes dark chapters of this fascinating past.

Orchids are more than just synecdoche for human sexuality, of course. Part One dives into that passionate world, but parts Two through Four examine distinctive moments in orchid science, business, and culture. Widening the story of orchids has revealed new and beautiful tales of human adventure and devotion to orchid cultivation over millennia. The less-known corners of the orchid world have not only given birth to feats of scientific observation and industrial innovation, but also inspired unlikely alliances across classes and cultures. They've driven humans into insanity and poverty, been a source of commerce and obsession, and inspired artistic works of stunning beauty and originality. Orchids are found on every continent except Antarctica, and orchid history, it turns out, is just as diverse.

Particular orchids focus the narrative of each chapter. Included for each flower—or a near relative better suited to current cultivation—is growing information (light, temperature, watering and fertilizing schedule, growing media, seasonal changes, special requirements) as well as features not usually included in orchid guides, such as country of origin, blooming season, duration of bloom, plant and flower size, as well as fragrance. As an indoor orchid grower for almost twenty years, I find these details crucial when deciding whether to purchase an orchid for cultivation and display at home.

Set together, the profiles gathered here are meant to deepen our understanding of the flower-mad and of the flowers that beguile us all. I hope your plants grow healthier and happier by virtue of appreciating the human dramas behind these blooms.

PART I

Orchid Love, Orchid Lust, and Orchid Sex

Chapter 1

S

Lusty Ladies of the Enlightenment

There are just some people in history you can't help but love. People whose prose and ideas still leap off the page, as invigorating and odd as when written more than two hundred years ago. Though he lived in an age of reason and experimental science, one prominent eighteenth-century English thinker believed that *lust*, beyond anything else, drove the life force not only of humans, but also of plants. Perhaps not coincidentally, he also had a passion for orchids. The flowers in his poems were so lively that they educated a wider public about the Linnaean system while simultaneously pushing natural history in new directions. His Georgian-era work is radical in its licentious focus on women's sexuality, the possibility of polyamory as a natural act, and the insistent revelation that *sex is all around us.*

Erasmus Darwin (1731–1802), grandfather of Charles, was a highly respected physician and inventor. He also published *The Botanic Garden* (1791), the most controversial prose poem collection of its decade. An atheist and evolutionist well before his time, he anthropomorphized orchids in it as threatening seductresses and comforting mothers. The work presents a parallel green universe—one similar to our own world but more amorous, more vibrant and threatening—where a "Botanic Muse" assembles her nymphs and collects exotic and indigenous blooms. Darwin's fantastic scenes portray plants masquerading

Portrait of Erasmus Darwin by Holl, after J. Rawlinson. Thornton, *New Illustration of the Sexual System of Carolus Von Linnaeus and the Temple of Flora, or Garden of Nature* (1799-1807).

as women from all walks of life and are rife with metaphors of orchids' feminine sexuality and power.

Darwin made the active choice to not live in London, giving up the very hub of Enlightenment scientific activity to pursue a country life. In 1778, he purchased eight acres of swamp near Lichfield, 130 miles northeast of London, and transformed it into a botanical pleasure garden—in part as an extended act of courtship for his second wife. Although so heavy that he had to cut a large notch out of his table to be able to sit and eat, Erasmus was an avid gardener. He chose his residence for its landscaping potential and its prospects over "pleasant and umbrageous fields." He terraced slopes near his house, mixing newly laid lawns with shrubs, lilacs, and rose bushes. There, he also enhanced a natural spring, created lakes, and crafted flower-lined paths.

His labors led him to publish *The Loves of the Plants* in 1789 (soon republished as the second prose poem in *The Botanic Garden*). Darwin initially chose not to put his name on it because he did not want the world to think it "a work of consequence" and believed it might pose a "professional danger" to his practice as a physician. He had other, far more serious tasks at hand, too. One such project was a full English translation of Linnaeus's complete works he was doing with local friends. Another was an attempt to modify continental weather patterns by advocating that European navies tow icebergs to the equator to moderate northern winters and cool the tropics. According to his grandson Charles, Erasmus need not have worried about *The Loves of the Plants*. His "proem" about plant procreation was instead met with "great and immediate" acclaim.

Crafted as an entertaining instructional guide to the Linnaean system for plants, *The Botanic Garden* is written in rhyming verse, with copious footnotes that ground Darwin's lofty couplets within specific plant species that exhibited Linnaean lessons best. *The Loves of the Plants* was explicit in its focus: plant procreation dictated most plant functions, as well as plant-insect relationships. Darwin did not hold back in his characterization of the carnal acts of flowers: orchids could be loving, violent, duplicitous, and sometimes downright murderous.

Plants were just as much engaged in "vegetable love" as they were in "vegetable adultery."

Darwin's poetry transformed gardens and jungles into magical worlds of sylphs, knights, alluring maidens, and angry gods. Its subtext hinted at the progressive political agenda that drove his scientific outlook: he was vehemently against slavery, had nothing but criticism for the church, and was an early supporter of the French Revolution. And, perhaps most of all, he approached the orchid family as one that was deeply symbolic for its overt sexuality and its potential for scientific and social progress. Darwin's feminine floral characters' attention focused on only one thing: themselves, their pleasure, and the continuation of their vegetable lineage. In his poetry, Darwin punctuated Linnaeus's new paradigm: flowers are organs of reproduction, full stop. And orchid women—god forbid—were active players in sexual acts both pleasurable and threatening.

Linnaeus himself had written of male and female flower parts decades before, assigning them the roles of "husbands" and "wives," with flower petals serving as their "marriage beds." A patriarch of his time, the early botanist based his taxonomy upon the "male" parts of plants with stamens defining Linnaeus' taxonomic classes; pistils, the "females," are secondary to them. But against Linnaeus, his beloved "Swedish Sage," Darwin reversed that priority. For Erasmus, female parts of flowers provided both the elements of reproduction and the location for affairs and gestation to take place. In his plant cosmology, the female took center stage.

Darwin includes orchids throughout *The Botanic Garden*. Animate players battle at cross purposes: a Chilean hummingbird robs various flowers of their nectar "treasure" throughout the jungle, but meets its match in a cypripedium orchid. As the small bird approaches the orchid, "Fair CYPREPEDIA with successful guile / Knits her smooth brow, extinguishes her smile." The flower's shape is a trap. "A Spider's bloated paunch and jointed arms / Hide her fine form, and mask her blushing charms;" its pouch could swallow the bird whole, but the "panting plunderer" decides to fly away.

In Darwin's recounting, what he calls a South American cypripedium has a "large globular nectary" that she must defend from hummingbirds. The plate included with the poem is a North American species, however, likely *Cypripedium reginae*, the showy lady's slipper. The pink and white spring bloomer can be found in Canada and the United States from North Dakota, east to the Atlantic, and Saskatchewan to Arkansas. Darwin explains that the "South American" cypripediums he references in the poem are larger and more colorful than their northern

relatives (also untrue—both continents boast large and vibrant cypripediums). The good doctor believed the entire genus sported cunning flowers with frighteningly spider-like inflated pouches, the orchids protecting themselves from small birds who would "plunder [their] honey."

Here, Darwin was mistaken in the mechanics of flower sex as well as its intent (which is predictable, since there's very little chance Darwin could have seen these orchids in person). Cypripediodiae do not have nectar, although some species use scent to lure their pollinators

Cypripedium
in Erasmus
Darwin,
*The Botanic
Garden* (1791).

(mostly bees, wasps, and flies; not hummingbirds). And unfortunately for their "panting plunderers," it is a rare orchid that rewards its pollinator. What's more, Darwin believed that what nectar was produced by orchids was meant to stimulate pistils and stamens into their lascivious work. Instead, most orchids produce not amorous juices but sophisticated lures for insects. Perhaps Darwin's Greek equivalent for cypripediums should have been the Sirens—here insects crash upon their orchidaceous shores, do their amorous and unwitting work, but live to tell the tale.

But there are several native English orchids Darwin did see, including brilliant insect mimics *Ophrys insectifera* (the fly orchid) and *Ophrys apifera* (the bee orchid). Darwin offers local detail of orchids in Derbyshire and the insects he thinks attacks them. He describes the fly orchid as resembling a small bee, so much so "that it may be easily mistaken for it at a small distance." He surmises that the orchid's similarity to bees means that "it is probable that by this means it may often escape being plundered." Again, Darwin had it backward: we now know, thanks in part to the more famous Darwin, that the bee orchid looks like a bee not to

Ophrys apifera in The Naturalist's Miscellany (1789).

scare away insects, but rather to invite a real bee (actually a small wasp) into copulating with it, thereby ensuring pollination. In a large *Cypripedium reginae* like the one Darwin dreams of, the big, deep pouch encourages the pollinator to stick around, affixing pollinia on the bee's thorax. The bee then crawls up the back side of the inflated pouch and exits under the staminode—a shield protecting the orchid's reproductive parts from the elements—ensuring cross-pollination when the bee visits other lady's slippers.

Regardless of how orchid pollination really works, for Erasmus, plant reproduction was the "chef d'oeuvre, the masterpiece of nature." Before we dismiss this sentiment as romantic claptrap, be aware that "masterpiece" in the eighteenth century was a popular euphemism for vagina. Increasingly unafraid to be bawdy, by the time his posthumously published *The Temple of Nature* came out in 1802, Darwin had honed his message to a fine point.

> Hence on green leaves the sexual Pleasures dwell,
> And Loves and Beauties crowd the blossom's bell;
> The wakeful Anther in his silken bed
> O'er the pleased Stigma bows his waxen head;
> With meeting lips and mingling smiles they sup
> Ambrosial dewdrops from the nectar'd cup;
> Or bouy'd in air the plumy Lover springs,
> And seeks his panting bride on Hymen-wings.

One gets the sense that Darwin might have been writing from his own experience. Twice happily married and with plenty of affairs in between—producing fourteen children—one of his obituarists cheekily wrote that Darwin "could never forsake the charms of Venus." In a footnote to the immodest stanza above, the poet notes the types of flowers that elaborately "protect their honey-cups" ("honey pot" and "honey cup" also being slang for vagina). Orchids, Darwin intoned, are one of those ladies who often frustratingly "complicate the apparatus" of coitus. (Perhaps our poor Erasmus had familiarity with this as well?)

Darwin's friend R. L. Edgeworth, an Anglo-Irish inventor and politician, wrote upon publication of *The Botanic Garden* that the poet should be placed among the likes of Homer, Milton, and Shakespeare in world literature. Darwin instead envisioned himself as the counterpart to Ovid, writing that the great Roman poet "did by art poetic transmute Men, Women, and even Gods and Goddesses, into Trees and Flowers; I have undertaken by similar art to restore some of them to their original animality, after having remained prisoners so long in their respective vegetable mansions." Darwin, in other words, was transforming plants back into men and women, gods and goddesses, and spun epic illicit tales to illustrate their dazzling lives.

Edgeworth in fact read many sections of the text to members of his family and attested that it "seized such hold of my imagination, that my blood thrilled back through my veins, and my hair broke the cementing of the friseur [hairstylist], to gain the attitude of horror." Darwin's poetry was so hair-raising, apparently, that his friend felt kissed by the twin emotions of the sublime: awe mixed with terror. Another English writer, Horace Walpole, deemed it "the most delicious poem on earth." Others weren't so impressed. Poet Samuel Taylor Coleridge explained simply that "I absolutely nauseate Darwin's poem."

Readers throughout time have simultaneously loved and raised an eyebrow at Darwin's *The Botanic Garden*. Misgivings about the work piled up, just as much as its sales receipts did. Religious leaders declared it a source of moral decay, where "lustful boys anatomize a plant." One concerned reverend was especially wary of women reading *The Loves of the Plants*—its representation of the sexual system putting "female modesty" at risk.

Darwin was in step with trends in his time—throughout the Enlightenment in general, sex was redefined as part of the nature of all living things. No longer a vice or sin, philosophers waxed that sex was

Emma Crewe, *Flora at Play with Cupid,* frontispiece to
The Loves of the Plants (1789).

simply a natural impulse, and should best find free expression in life.
As such, just as Linnaeus's sexual system for plants was taking hold, so
too were a glut of pornographic texts published, men with mistresses
somewhat normalized, and children born out of wedlock more often
accepted within extended families.

But Darwin had clearly struck a sensitive nerve. Even the frontispiece
"Flora at Play with Cupid," drawn by Emma Crewe, was thought beyond

Jan Brueghel the Elder and Peter Paul Rubens, *Flora and Zephyr* (1617).

the pale. Critics protested it had "an air of voluptuousness too luxuriously melting." As one might expect, the image suggests much more
than a goddess watching a cherub. More seventeenth-century euphe-
misms for female sex organs, and the vagina in particular, included all
possible variants of "Cupid's cave," "Cupid's hotel," "Cupid's nest," and

"Cupid's ring." Add to these suggestive metaphors Enlightenment code words for the general female sexual apparatus: a woman's private parts were a "garden," "hortus," "*Frutex vulvaria*," "flowering shrub," and "fruit shop." Darwin here relentlessly underlined—for ladies as much as for men—multiple popular ribald meanings for a woman's "mouth of nature," "secret of nature," "paradise," and "mystery," to be sure.

Darwin's readers would have also been aware of Flora's checkered past in Roman mythology. Before she was included in the pantheon as a goddess, she was an infamous courtesan. Upon her death, she left much of her illicit gains to the city of Rome and was deemed the deity of flowers and protector of prostitutes. Highly skilled, Flora regularly sold herself to the highest bidder, often bankrupting her lovers with increasing demands in payment for her attentions. The goddess was not reformed upon entry into the heavens, either: she married Zephyr but soon took up with Hercules.

Less racily, Darwin's poems linked gardening and botanical study as popular leisure activities for both men and women in the late eighteenth century. What was more, the fashion for gardening easily transferred to something as far afield as women's dresses and extravagant hairstyles of the late 1700s. When a group of well-dressed ladies came to visit, English playwright and moralist Hannah More scolded: "I hardly do them any justice when I pronounce that they had amongst them, on their heads, an acre and a half of shrubbery, besides slopes, garden plots, tulip beds, clumps of peonies, kitchen gardens, and greenhouses." Although More exaggerates, English women did follow the French trend in wearing their hair very high and ornamented with feathers, flowers and greenery, butterflies, caterpillars, beads, and blown glass. Flora's luxuriously flower-spangled coiffure in Darwin's frontispiece was well in style after all. Humorists like Matthew Darly satirized the trend in the 1770s and '80s, sketching contemporary English hairstyles to include not just vines, flowers, and manicured trees, but full parterres managed by miniature gardeners.

But not all orchids were lewd in Erasmus' daydreams. At home in England, Darwin found terrestrial orchids in his own wet meadows to be perfectly maternal. Writing about *Orchis morio* (now *Anacamptis morio*), the poet personifies its life cycle and gives human meaning to its underground tubers. For this orchid, last season's tuber shrivels to feed the current year's growth.

> With blushes bright as morn fair Orchis charms,
> And lulls her infant in her fondling arms;
> Soft plays of *Affection* round her bosom's throne,
> And guards his life, forgetful of her own.

Matthew Darly,
*The Flower
Garden* (1777).

THE FLOWER GARDEN.

The purple *Orchis morio*, commonly known as the "green-winged orchid," is found in central-southern England, Wales, and Ireland, where it blooms in grassy meadows from late April to June. Darwin had firsthand experience with the orchid—he describes the "insipid mucilaginous taste" of its boiled tubers, and in fact attests to having seen "the *Orchis morio* in the circumstance of the parent-root shriveling up and dying, as the young one increases." The mother sacrifices herself for her child—Darwin found this analogous to other tubers, corms, and bulbs he was familiar with, especially the tulip. And so his local *Orchis morio* is depicted as the perfect, self-sacrificing mother. Orchids from tropical regions, however, are anything but.

One of the most chilling moments in *The Botanic Garden* includes

Orchis longicornu here, Orchis morio for Darwin, now Anacamptis morio. Robert Sweet, *The British Flower Garden* (1823–1829).

an orchid only described as "epidendrum." In Darwin's day, epidendrums were defined as any epiphytic orchid. Epiphytes—plants that grow on trees but do not steal nutrients from them—were erroneously lumped with all other parasites, including dodders and mistletoe (actual parasites), as well as ivy, clematis, tillandsia, mosses, and fungi (none of which are parasites). The poet describes feminized epidendrums as crawling their way through the air, awaiting the moment they can latch onto their next host:

Two Harlot-Nymphs . . .
With labour'd negligence, and studied ease
In the meek garb of modest worth disguised
The eye averted, and the smile chastised
With fly approach they spread their dangerous charms
And round their victim wind their wiry arms.

Although they may look like flowers in "meek garb," parasitic plants, the epidendrum harlot-nymphs, are here the dreaded serpents of Greek myth. Although Darwin was smitten with women in his own time, he felt free to import ancient misogynistic tropes. Darwin continues, "The scaly monsters roll'd / ring above ring, in many tangled fold / Close and more close their writhing limbs surround / And fix with foamy teeth the envenom'd wound." Trees and shrubs, cast as male, bear the female orchid's assault stoically, while "silent agony sustains their rage." The once-seductive orchid is now additionally poisonous. "The harlot waves in the air and enwraps its victim, encouraging the tree to 'drink deep, and forget your care.'" Not to be misunderstood, Darwin surmised that these parasites not only fed off of their unwilling hosts, but ultimately strangled them in the bargain. Some of these plants, Darwin explained, served to "incommode the taller trees." Epidendrums (all epiphytic orchids), he explained, "incommode them all"—tall trees, shrubs, and anything else they attached themselves to.

We may wonder what "epidendrum" Darwin was referring to,

describing an orchid with powers the likes of the Greek sea monster Scylla. The poet wouldn't have had many orchids to choose from. The first edition of *Hortus Kewensis* (1768), a list of plants the Royal Botanic Gardens at Kew had in their collection, contained only twenty-four orchid species. Most were British natives; only two were non-European. Of the two, the orchid Darwin likely found threatening was *Epidendrum retusum* (today *Rhynchostylis retusa*, from India and China). It has thick, wiry roots that extend away from the epiphytic orchid, seeking to stabilize itself upon larger sections of tree trunk. When blooming, the large rhynchostylis and other similar epiphytes also exude a sweet, sticky sap to tempt their pollinators. Darwin could have mistaken its fluid as venom harmful to trees.

If Darwin thought the flower, not the roots, looked like arms enwrapping a victim, the most likely culprit was *Prosthechea fragrans* from Central America. The third volume of *Hortus Kewensis*, published in 1789, contained forty species, but still just fifteen non-British species of orchids appear. Key among them, the first tropical epiphyte had bloomed at Kew just seven years earlier. *Prosthechea fragrans* opened in Kew's hothouse in October of 1782, its honey scent enveloping all passersby. Known as the cockleshell or clamshell orchid, it sports a candy-striped lip and resembles a small dove in flight. Five years later, a cockleshell orchid in the same genus, *Prosthechea cochleata*, bloomed there as well.

Prosthechea fragrans as *Epidendrum fragrans* in *The Botanist's Repository* (1810).

Kew took an early interest in orchids in part because they were the most exotic-looking plants sent back to England as the British Empire was expanding around the globe. By 1815, the empire held territory in North and South America, Africa, India, and Australia. The hothouse's name, the Great Stove, was apt: eighteenth- and early nineteenth-century gardeners tended to run their greenhouses far too hot and wet all year long; most species were nearly boiled alive. The orchids that lived more than a year in these conditions were hardy indeed.

With orchid shipments to Britain increasing every year and general knowledge of orchids on the rise, botanists and horticulturists took note. Among them, Sir William Jackson Hooker grew orchids at his home before he became a professor of botany in 1820. When he was appointed the first director of Kew in 1841, he immediately acted to increase its orchid collection. He was equally as interested in less-showy orchids as their more dramatic kin, which were the current rage of the Royal Horticultural Society and anyone with money to buy them. Kew soon received not only orchids from expeditions abroad, but also large bequests from wealthy families. The botanical garden then built more greenhouses, including cooler "orchid pits" in the mid-nineteenth century, where the plants thrived. Kew's early focus on and ultimate success with orchids was part of the tinder that eventually exploded into full-blown orchidomania across Europe and America. Joseph Dalton Hooker succeed his father as Kew's director in 1865, and only intensified the botanical garden's focus on orchids. Kew soon grew to house dozens of genera, including cattleyas, oncidiums, dend-robiums, lycastes, brassavolas, cymbidiums, and aerides.

While Darwin was inspired by the early orchids at Kew, ultimately, *The Botanic Garden* and his other works also set the tone for how Westerners would conceive of orchids for the next century. The flowers were beguiling women to seduce, to endure, and be threatened by. As such, while setting the female at the center of his plant cosmology, Darwin also amplified stereotypes of women in the eighteenth- and nineteenth-century English imagination, including those in the countries taken

Flora Attired by the Elements, designed
by Henry Fuseli and engraved by Anker
Smith, frontispiece to *The Botanic
Garden* (1792).

over by the British. Local orchids were chaste women protecting their honey-cups, and loving, self-sacrificing mothers. Foreign orchids from Indochina and Central America were instead alluring and poisonous parasites. Darwin had turned a key during the Enlightenment—it unlocked a door into a garden where no one could soon forget that flowers have sex lives. And for as much as Darwin himself enjoyed encounters with Venus and Flora, their flowers could be dangerous. Orchids, in particular, were dreams and nightmares brought to life, harkening back to ancient times and driving nature into a strange new sexual future.

PROSTHECHEA FRAGRANS

Prosthecheas arguably did more than any other genus to ignite the orchid craze in Europe. Also known as octopus orchids, their labella tilt to the sky like a queen's crown, while their thin petals float below it like the legs of an elegant cephalopod. Both bear the scent of expensive perfume; *fragrans* is light and sweet, whereas *cochleata* is richer and more complex. Their non-resupinate flowers (lip on top, petals and sepals below) bloom sequentially on the same stems for many months. Cockleshell orchids' mix of curious features along with prolific blooms—their swollen pseudobulbs at the base of their leaves fat with stored nutrients—will wrap their wild tendrils around any anthophile's heart.

Orchid Details

— Place of origin: Mexico, Central America, northern South America
— Blooming season: year round
— Flowers last for: several weeks; a single inflorescence can keep producing blooms for years
— Plant size: 1.5–2 feet tall from pseudobulb to leaf tip; flowers held clear of leaves
— Flower size: 1.5–2 inches across; several can bloom simultaneously per spike
— Fragrance: sweet honey, vanilla, gardenia
— Plant habit: upright, arching, sympodial orchid with clumping pseudobulbs

Orchid Needs

— Light: moderate to bright, no direct sun
— Temperature range: in winter, low 50s to low 70s°F; in summer, low 60s to low 80s°F
— Humidity: 60%–80% ideal; will usually cope with less
— Water: quite high May through October (up to 7 inches per month); much lower rest of year, and nearly dry February and March
— Growing medium: fine to medium fir bark with perlite, can also be mounted

— Fertilizing schedule: weekly one-half dose balanced orchid
fertilizer when growing

— Seasonal changes: keep orchid cooler and rather dry in
winter, but keep relative humidity up

— Special requirements: roots must dry rapidly, and be com-
pletely dry before watering again

Chapter 2

❦

Orchids Fit for a Chinese Empress

C hina has, in all ways one might imagine, the world's most elaborate culture of flowers. Its story—much like the story of the
empress at the center of this chapter—is marked by delicate
beauty, botanical wealth, the trade in elite concubines, and ruthless
imperialism. China's staggeringly deep garden history is only topped
by its bewildering natural riches. Relative to Western forms of refined
horticulture, theirs is a tradition spanning millennia, not centuries.

We know that Chinese gardening in the form of cultivating green
retreats and as a restful pastime is at least as old as the Six Dynas-
ties (220–581 AD). But scientists and historians agree that the cultural
appreciation and sometime cultivation of key flowers in China began
far earlier. Peach and plum blossoms, camellias, chrysanthemums,

lotus, peonies, and cymbidium orchids have histories in Chinese culture extending to at least 1000 BC. Since the fourth century BC, various
orchids were also an essential part of Chinese pleasure gardens—delights
to the eyes, ears, and nose. Their owners walked through galleries of flowers, often with the sole purpose of catching a cymbidium's scent on the
breeze. For centuries, continuing to the present, important plants were
thought to embody strong principles and give form to human emotional
states. Orchid appreciation was both an aesthetic and moral experience.

Empress Dowager Cixi (1835–1908) was especially fond of cymbidium

BOTANIQUE.

MONOCOTYLÉDONES.   Orchidées. (June)

Turpin pinx.t et direx.t   Massard sculp.t

ANGREC de Chine.
EPIDENDRUM sinense. (Andr.)
(¹/₃ Grand. nat.)

1. Pistil et etamines. a. Ovaire. b. Style et trois etamines soudés en un seul corps.
c. Etamines rudimentaires.d. Etamine antherifere. e. Stigmate. 2.Fruit coupé hori-
zontalem.t 3.Graines.4.Id grossies.5.Une graine dont on a enlevé une port.n de l'arille. a.

orchids. Since ancient times,
cymbidiums had been called
the "first fragrance" and the
"king of fragrance" with
their warm, rich, and intoxicating scent. Their aroma
was described as "discreet
yet interesting, subtle yet
pervasive . . . like the friendship of an honorable man."
In roughly 500 BC, Confu-
cius lectured that cymbidium orchids were a model for
all *junzi* (virtuous people),
because they were temper-
ate flowers, growing in quiet
valleys, yet they bettered the

Fragrant *Cymbidium sinense*
(here *Epidendrum sinense*)
blooms during Chinese
New Year. *Dictionnaire
des sciences naturelles*
(1816–1830).

Katharine Carl, *The Empress Dowager, Tze Hsi, of China* (1903).

world by exuding a luscious bouquet. Huang T-ing-chien, a Sung Dynasty poet, declared cymbidiums the "national fragrance" and waxed on about their "perfection." It would seem surprising that Cixi, a woman so appreciative of a delicate-seeming flower, would be called "dragon lady" and be a master of political moves.

Cixi is today regarded as the most powerful woman in Chinese history. In 1851, at age 16, she was presented at the Chinese Emperor's palace, which was secluded deep within Beijing's Forbidden City. Given the name of Lan—"Orchid"—upon her entry at court, she became one of Emperor Xianfeng's favorite concubines and bore him his only son, Zaichun. While she climbed the ranks as a consort, she was given additional names meaning "exemplary" and "virtuous." When Zaichun was five, Emperor Xianfeng died and Lan was transformed into the queen regent nearly overnight. A smooth politician, she pulled off a coup, upended the rules of succession, claimed the title of Empress Dowager Cixi (meaning kindly, joyous, motherly, and auspicious), and outwitted many powerful heads of state.

Despite all of her honorifics, during her reign, she actually preferred to be called "Father." As a woman and previous courtesan, she clearly had many gender barriers to topple to lead an empire. Being called "Father" would have reminded all men that came to formally consult her—Cixi always behind a screen, as women were not to be seen in the halls of government, let alone have anything to do with state affairs— that she held ultimate authority.

Early in her rule the empress warned: "Whoever makes me unhappy for a day, I will make suffer a lifetime." She had, at twenty-five years old, for example—in the coup that made her empress dowager—executed one minister and strongly suggested suicide to two others (they accepted their fate). Decades later, she imprisoned her stepson for attempting to assassinate her; years after that, she poisoned him because she felt he was weak and influenced by the wrong advisers. Cixi retained her steel nerves, as well as her bravado, until the end of her 47-year reign: "I have often thought that I am the most clever woman that ever lived, and others cannot compare with me . . . Although I have heard much about Queen Victoria . . . I don't think her life was half so interesting and eventful as mine." By the end of her tenure, she had outlawed foot-binding, expanded the freedom of the press, sanctioned intermarriage between Han and Manchu people (ethnic rivals for centuries), withstood the multi-pronged Boxer Rebellion, transformed China into a voting constitutional monarchy, and modernized the country in no small way.

The garden-inspired aesthetic choices of Cixi were as strong as her political acumen. She cultivated orchid gardens on the Imperial Palace grounds and at the Summer Palace ten miles outside of central Beijing. Filling their halls with porcelain jardinières of potted and cut orchids, she also topped her coiffure with fresh flowers brought in daily from the countryside. Courtiers tried to gain favor by sending in special flowers from their estate gardens. As part of engineering China's image to the world, she broadcast dozens of portraits and photographs made of her. In most she is surrounded by plants and flowers, her robes also woven with botanical themes.

The empress often appeared in a yellow satin gown, embroidered with gold cymbidiums and pink peonies, and an elaborate jade and pearl headdress in the shape of a phoenix and flowers. One of Cixi's portraitists later in life, American painter Katherine Carl, made a point of including orchids in those works. She enthused about the orchids and noted Cixi's special love for cymbidiums: the empress's palaces had "flowers everywhere! [There] bloomed a sort of orchid, of delicious fragrance, of which her majesty is very fond . . . the combined odor of fruits and flowers gave a subtle, composite perfume quite indescribable and delightful."

One of the orchids Cixi would have had ample access to is *Cymbidium ensifolium.* Cymbidiums today remain one of four botanical Chinese

Cymbidium ensifolium as Epidendrum ensifolium. Zorn, Auswahl schöner und seltener Gewächse als eine Fortsetzung der Amerikanischen Gewächse (1795–1798).

Epidendrum ensifolium.

"wisemen" or "gentlemen," taking their place with the plum tree, bamboo, and chrysanthemum as revered botanical symbols that also represented the seasons (plum blossom for winter, cymbidiums for spring, bamboo for summer, and chrysanthemum for fall). *Cymbidium ensifolium* is a petite orchid with a wide range of color tones. Perfect for the garden or the windowsill, it has elegant and lance-like leaves (*ensifolium* means having sword-shaped leaves). Its petals are often striated and its lip is speckled, with blooms naturally appearing in rich yellows, greens, maroons, pinks, and browns.

The empress dowager's court was lavish, and other flowers were present as well. Living blossoms and plants that were depicted in ancient scrolls and paintings surrounded Cixi. Camellia motifs appear in her clothing and palace decorations, the flower being long revered in China not only for its elegance, but intelligence, because of its long blooming period. Like pine and cypress trees, the Chinese thought it displayed nobility and virtue. Cixi also loved peonies; their large, fragrant blossoms had been known as the "king of flowers" since the Tang dynasty (618–907). And since the 1750s, the white moth orchid *Phalaenopsis amabilis* had joined the ranks of China's favorite orchids. But Cixi did not favor them and their rigid, staked flower stems, despite the orchid embodying good luck in business and love, its common name being "Empress among orchids." Perhaps Cixi didn't want the competition from another "empress."

Cixi both followed and set flower trends. By the mid-nineteenth century, many Chinese women decorated their hair and clothing with fresh or artificial flowers. Women of different classes would use the blooms they had access to: country women might use red hibiscus, while women in cities bought magnolias and gardenias sold by street vendors. Brighter colors became associated with lower classes, and paler shades and more delicate flowers with the upper ranks. Yet women in the empress dowager's entourage were not allowed to wear fresh flowers unless given to them by Cixi— she wanted the distinction for herself and thought that fresh flowers better matched older women's countenances. The empress also mixed the oils of jasmine, peony, and cymbidium orchids to create her own perfumes.

Empress Dowager
Cixi and Princess Der
Ling with attendant
on Peony Mountain
(c. 1903–1905).

A devout nature lover, Cixi had a strong sense of design about her many palace gardens. Their long vistas, artificial lakes, and framed views seemed to startle visitors with their grandiosity. The empress had hundreds of gardeners—often a duty for eunuchs—at her disposal and would direct their activities. She transformed one garden into a working farm with pear and apple trees, chickens, and crops. At times, she trimmed trees, planted bulbs and flowers, and collected blooms. Each year, Cixi led the court ladies in taking chrysanthemum cuttings and propagating them in flowerpots. Eventually the empress ordered an impressive "peony mountain" built; visitors reported that in the month of June, Cixi's imperial gardens were a solid mass of flowers and foliage. Touring her gardens was a favorite everyday pastime, as well as a show for state visitors. When her friend Princess Der Ling first arrived at court in 1903, she said that Cixi "walk[ed] along the long veranda, so we followed her. She showed me her flowers, and said that she had planted them herself." Cixi expected floriferous gifts, and also bestowed them: "every now and then potted peonies and orchids from her gardens, baskets of fruits from her orchards, boxes of cakes and balls of tea would arrive at the legations, bearing Cixi's good wishes."

Beyond the palace walls, orchid gardens were found throughout China. Famed Chinese author and painter Su Hua (Ling Shuhua), recalling her childhood at the turn of the twentieth century, recorded that there were more than twenty kinds of orchids in her family's greenhouse in Beijing. Nearly all of them were endemic to southern China, several hundred miles from the capital. The most prized was the *Chien lan* (*Jian lan* or Fujian orchid, known in the West as *Cymbidium ensifolium*), with its delicious scent and variable color, especially loved by Hua for its light-green flowers and hearts. Rare orchids in their collection also included the *Ouk-lan* (*Yue lan* or Guangdong orchid). Shandong orchids, with their green petals and red hearts—from the eastern peninsula—were common in all greenhouses of the capital. And many orchid lovers possessed an unnamed lavender spray orchid that had been imported from Sichuan in the west. Caring for orchids was part of a garden's seasonal cycle. Rather ingeniously, Hua, along with the family gardener and his friend—a former gardener at Cixi's Summer Palace—would collect horse hoof parings and steep them in a broth to feed the orchids. Horse hooves contain phosphorous and calcium, essential macro- and micronutrients. Applied in early spring, the phosphorous in their "orchid soup" enhanced the orchids' root systems; phosphorous is still today considered a major "bloom booster" when applied consistently through the year or in higher amounts just before the bloom cycle. Cymbidiums' scent added yet more texture to the sensorial pleasures of urban gardens.

In Cixi's era scent was highly valued in part because like most nineteenth-century cities, Beijing often declared itself by its stench. Sterling Seagrave, a biographer of Cixi, describes the capital's ubiquitous stink: "Out of the Forbidden City and through the legation quarter ran an

Tapestry of cymbidiums, other flowers, a rock, birds, and butterflies, Qing dynasty.

open sewer that perfumed the air night and day, lending its unmistakable scent to tennis matches, soirees, and formal dinners," and penetrating sleeping areas at night. The malodorous miasmas of large cities were perhaps another reason why cymbidiums remained paramount in Chinese culture throughout the period.

Scent, of course, adds to the voluptuous nature of orchids. And orchids in particular have long been associated with Chinese courtesans and prostitutes. Centuries before Cixi's rule, women were given floral names, and courtesans of the royal court, like Cixi when she joined it, were each matched with a specific rank and flower upon entry into official court life. Male court painters favored the orchid as a subject, setting it in contrast to a solid and immoveable object, such as a rock. This tableau became a metaphor for a lovely girl in seclusion—the eroticism, mutability, and continuity of life and its pleasures. By the 1600s, courtesans also began painting orchids. While other flowers and trees were also acceptable subjects, the delicacy and refinement of the orchid was thought to be best aligned with women's artistic merits.

The Mustard Seed Garden, a revered Chinese painting manual since 1679, includes an entire chapter on painting cymbidiums. It instructs: "Leaves are painted in a few strokes, and they should have a floating grace in rhythm with the wind, moving like a goddess in a rainbow-hued skirt with a moon-shaped jade ornament swinging from her belt." Students are still taught to paint orchids when they are in a light mood, for the leaves and buds reflect one's happiness in their fluttering grace.

Yet in contrast to the West's incessant focus on blooms, orchid representation in Chinese culture was never centrally about the flower. Orchid painting, since its earliest days in China and in women's hands, was focused on the lithe leaves of the cymbidium—arching gracefully across a canvas. The flower was considered the orchid's heart, and later its eyes; painters were instructed to dot the heart of the cymbidium flower as one last flourish. Cymbidiums simultaneously held the romance of beautiful women, the power of great men, and the entangled "friendships" between the two. One might think that courtesans painting orchids across centuries would sully the symbolic weight of

Cymbidium orchid. *Eight Designs of Flowers*, formerly attributed to Huang Jucai (after 933), possibly Ming dynasty.

the cymbidium's place among the "four gentlemen" of revered plants in Chinese culture. Yet just the opposite is the case: like Cixi, other women had been presented to court as concubines but rose to power and respectability, and the orchid stoically retained its importance.

The rhythm of Cixi's court was set to the garden rituals practiced throughout the lunar calendar. The empress's sixtieth birthday in 1887 occurred at a moment when the empire was tottering. Her birthday still elicited lavish gifts, including money that poured in from rich merchants seeking honors. Cixi also raised substantial loans from abroad and at home, promising to equip and modernize the Chinese navy and naval academy. Instead, at the Summer Palace, she built elaborate courtyards and halls for a mile-long garden, which stretched along the northern shore of Kunming Lake.

Roundly critiqued for prioritizing her gardens above all else, historians have accused Cixi of showing her plants consideration and love that she did not show to human beings. Taken in the broader context, we might

Empress Dowager Cixi and attendants on Lotus Lake inside the Forbidden City (c. 1903–1905).

wonder whether focusing on the empire's gardens was instead the choice of a strong woman standing up to the crushing sexism of her age. More likely, it stemmed from Cixi's first impressions of court life and the atmosphere of the Forbidden City when she ascended the throne. For the Old Summer Palace was utterly destroyed in 1860, just one year before Cixi declared herself empress dowager. British and French forces had battled their way into the Forbidden City, acting on joint state orders to forcibly open China to trade. They looted the palace, and then set fire to it. Historian Jung Chang writes, "The fire, fueled by more than 200 opulent and exquisite palaces, pavilions, temples, pagodas and landscaped gardens, raged for days, enveloping west Beijing in black and ashen smoke." It was not just the palace and riches that were sacked and burned, but also plants and trees, some that were centuries old, and most cultivated through multiple generations of rulers and gardeners. British Lieutenant Colonel G. J. Wolseley wrote of his brutal deeds there with Lord Elgin: "When we first entered the gardens they reminded one of those magic grounds described in fairy tales; we marched from them upon the 19th October, leaving them a dreary waste of ruined nothings."

Over ensuing decades Cixi retaliated, and China restabilized during her long rule. Importantly, author Qian Xingjian has described how cymbidium orchids under Cixi's reign and into the modern era took on new meaning for Chinese people, writing:

> When our motherland was weak or disgraced, the orchid would become the vehicle and symbol for patriots to express their national integrity. While painting orchids, some artists would expose the roots uncovered by mud, so as to express their indignation about the national territory lost or ceded in their times. People love, admire, cultivate and eulogize the orchid, because it can offer an aesthetic pleasure, spiritual comfort and enjoyment.

American painter Katherine Carl underlined that Cixi herself was still comforted by flowers at the turn into the twentieth century: "She

Katharine Carl, *The
Portrait of the Qing
Dynasty Cixi Imperial
Dowager Empress of
China by an Imperial
Painter* (1904).

had flowers always about her . . . however careworn she might be, she always seemed to find solace in flowers!"

Cymbidium appreciation has only grown in China and wider Asia. Today, Ten Shin Gardens, a major nursery in Taiwan and global supplier of orchids, details twenty-seven different cymbidium flower shapes, their descriptions ranging from flowers that look like the petals of narcissus, plum, and lotus, to resembling helmets, swords, lions, and butterflies. But more detailed by far are the distinctions between cymbidium leaves and the fine points of their variegation. Here, leaves vary from an ombre sunrise to patterned stripes, as well as claws, cranes, and cloudy fountains. Most are on offer with all levels of saturation in green, yellow, and white. Astoundingly, there are more than one hundred described types of cymbidium leaves in circulation.

The empress dowager died of stroke in 1908 at the age of seventy-two. Her gardens remain some of China's most visited attractions, and her

Summer Palace gardens, in particular, are now a UNESCO World Heritage Site. Cymbidiums' multifaceted place in Chinese culture had run parallel to Cixi's changing roles within the Forbidden City. Both were beautiful commodities, yet also meaningful actors, symbols of both China's adherence to tradition and its hard-fought change. Both Cixi and the orchids she loved were prized for their delicate yet immutable nature and ability to put on a show. The flower and the woman were simultaneously ethereal yet enduring—courtesans turned empresses.

CYMBIDIUM ENSIFOLIUM

Cymbidium ensifolium has a scent best described as a mix of jasmine, lily of the valley, and lemon. Cixi was known to wear its scent, and today several high-end perfumes derive their fragrance from these blooms. Orchids' aromas have gained more importance over time. Since 1989, the World Orchid Conference has included scent as a category for competition. Judged by professional perfumers and wine tasters, four qualities are considered: "intensity and diffusiveness," "elegance," "gorgeousness," and "liveliness and freshness." Most orchid growers today are well aware that buyers have an appreciation of orchids' fragrance and note that in their descriptions. Whether

you'd like to grace your table with an orchid that smells like roses, jasmine, coconut, root beer, cinnamon, or dog poop, there are more than enough to choose from.

Orchid Details

— Place of origin: Southern and Eastern China, Sri Lanka, Southern India
— Blooming season: summer through fall
— Bloom duration: two to three weeks (its strong fragrance shortens its bloom time relative to other cymbidiums)
— Plant size: 1.5 feet tall from pseudobulb to leaf tip; flowers usually held clear of leaves
— Flower size: 1.5 to 3 inches across; five to nine flowers per spike
— Fragrance: jasmine, lily of the valley, lemon
— Plant habit: sympodial, upright, arching, clumping pseudobulbs

Orchid Needs

— Light: moderate to bright, but no direct sun
— Temperature range: winter 50s–70s°F; summer 60s to high 80s°F; plants have been known to adapt to temps 6°F–8°F cooler as well
— Humidity: 70%–85% summer and fall; 60% winter and spring
— Water: high from June through August; lower rest of year, and nearly dry December and January
— Growing medium: fine to medium fir bark with perlite
— Fertilizing schedule: weekly one-quarter to one-half dose balanced fertilizer when growing
— Seasonal changes: keep cool and rather dry in winter
— Special requirements: instead of watering the roots in winter, try misting the leaves in the morning

Chapter 3

✑

Orchids in the Tenderloin

W hen I began searching for the first public orchid show in the United States, I thought I would encounter one with the usual nineteenth-century botanical specimens—which is to say grizzled, one-armed Amazonian explorers and haughty society ladies, those overdressed Victorian anthophiles with money to burn. I crossed my fingers for a few inventive floral designers, and I knew I would encounter some inquiring, bespectacled scientists. And I did find all of these personalities, and more. But what took me by surprise was *where* the first orchid show took place. Rather than in a botanical garden or a civic pavilion, a college campus or a seller's greenhouse, it turns out that the first orchid show in America was hosted by New York City's Eden Musée, of all places, a dime museum located in the heart of late-nineteenth-century Manhattan's Tenderloin District, one of the most notorious neighborhoods in all of Gilded Age America.

Dime museums have a storied history. Master showman P. T. Barnum established the first, which he called the American Museum, in New York City in the 1840s. Unlike the august institutions that sought to shape civic culture in previous decades, dime museums promised to

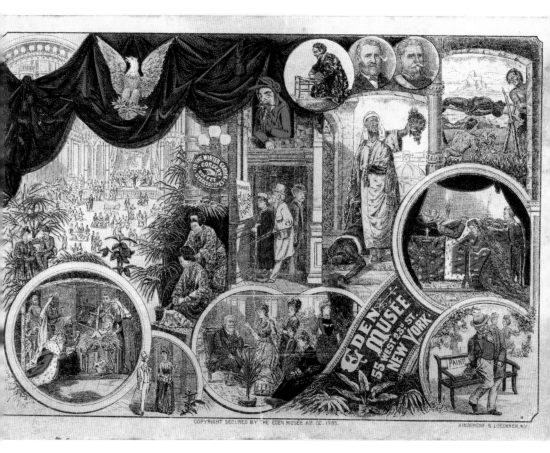

Eden Musée promotional card c. 1885, depicting the Winter Garden and several wax tableaus.

bring education and entertainment to the middle and lower classes, and for the first time consciously catered to women and children. These "museums" housed historical tableaus along with freak shows and waxworks, menageries, panoramas, melodramas, sideshows, and all manner of convincing fakes and oddities. Most were permanent small-scale circus-carnivals with rotating attractions, housed under one roof, central to foot traffic in downtown areas. Today, we stagger at the thought of gawking at humans with rare biological conditions paraded on stage, but dime museums of all stripes soon became a staple of large cities, promoting particular ideas—often negative when not white, elite, and American—about race, class, national identity, and America's global reach as it forged a dynamic new mass culture.

The Eden Musée (pronounced Moosey or Musey, because the hard-knock locals didn't put on airs) first opened in 1884. The museum's owners promised that they were running a "strictly first-class resort, particularly for ladies and children"—yet in the same breath boasted of a "fascinating assortment of horrors." This part of town (which today straddles Chelsea and the Flatiron District) was a single man's or straying husband's dream. Located on West 23rd Street between 5th and 6th Avenues, the musée was near the brothels at 29th Street, high-end gambling halls at 28th, and low-end gambling places even closer still at 27th. Saloons raged through the night on every corner, and bordellos could be found on every block.

By the 1890s, a night court was established to deal with the Tenderloin's scores of sex-trafficking cases. One month after opening, Eden Musée's manager was arrested for conducting business on Sundays. And by the summer of 1886, the owners were dragged into court again for failing to secure a license to stage concerts and minstrelsy. For as much high culture as it promised, the musée was run very much in the spirit of establishments throughout the Tenderloin—crudely, and willing to cut corners.

And so, orchids were launched onto the wider American scene within a magpie's nest of what were considered oddities at the time— our favorite flowers taking stage in an area of town also known

Siebrecht & Wadley catalog
(1891).

as Satan's Circus. Orchids might at first seem an odd choice for the
musée's promoters, pale competition for its tarot readings, serpentine
dancers, fire eaters, sword swallowers, Japanese jugglers, and Viennese
lady fencers. But remember that orchids, too, were novelties for Americans at the time—and what mesmerizing, delicate little monsters they
were. As patrons inspected the flowers, they described seeing solemn
owls and gaudy insects peering back at them. Still others were shocked
to find "waxlike human faces, recalling the dreams of elfland" in the
orchids' calices. The musée showcased operatic "red and bright colored
orchids, appearing like Wagner's dragons"—and visitors reported
that they were afraid the mythical creatures "would bellow forth ugly
sounds" at any moment. Perhaps it was not such a far jump from one
type of spectacle to this floral one. Regardless, the musée's first week-
long orchid show—mounted in March of 1887—was a colossal success.

Whatever their familiarity with these exotic blooms beforehand,
the *New York Times* wrote that visitors left the show "confirmed

SKETCHES AT THE ORCHID SHOW.—THE CAVE.

The Orchid
Cave at the
Eden Musée
orchid show.
*The American
Florist* (1887).

monomaniacs on orchids in general." And why not? New York horticultural firm Siebrecht & Wadley had turned the musée into a space
of "floral enchantment" and brought with them several other professional plantsmen toting their own floral spectacles. Late winter
visitors had stepped off of the city's grimy sidewalks and into the luxuriant tropics through a specially designed Orchid Cave. The show
offered banks of slipper orchids that "gleamed like green polished
moccasins," as well as "velvet cups of fluted cattleyas." In other corners, light and airy blossoms waved in the air, creating a "bright and
natural" scene. For the price of a 25-cent ticket (about $7 today), the
orchid exhibition transported patrons into an emerald universe of
cultivated delights.

The orchid craze that came to midtown Manhattan included other
tropical plants. Coconut palms offered a tropical backdrop to the

CENTRAL ARRANGEMENT AT THE RECENT ORCHID SHOW AT NEW YORK.

Just one of the displays at the 1889 Eden Musée orchid show. *The American Florist* (1889).

shows, and around them were staged "saucy-looking lavender-eyed" orchids. In its second year, the orchid show featured Japanese jugglers at the entrance, and one floral designer's pet monkey frolicking in the greenery. The musée also played up its international and mystical pretenses. When needing a break, men relaxed in the Turkish smoking room. The newly christened Winter Garden (a.k.a. the museum's concert hall) boasted skylights and French mirrors; extending a full city block, it was filled with orchids on display but could hold one thousand people comfortably when empty. At the far end of the Winter Garden, the museum's famed house band, its Hungarian Orchestra, played continuously during the exhibit—their "Orchid Show March" was arranged especially for the occasion.

In the gallery above, dozens of stereopticons provided views of "beauties of the human form." Wax tableaus throughout the building recreated Roman gladiators, famous artists and actors, great musicians, and rulers of the world. The musée focused on recent and long-past historical scenes, including Queen Isabella receiving Christopher Columbus after his "discovery" of America in 1493; US General Washington crossing the Delaware in 1776 (modeled after Emanuel Leutze's 1851 painting); French Emperor Napoléon III lying in state in 1873 with Empress Eugénie kneeling in mourning by his side; US General Custer's death at the Battle of Little Big Horn in 1876; and Chicago's Haymarket anarchists of 1886 meeting in secret. They also presented more fanciful life-size dioramas, such as Shakespeare's dramatic balcony scene in *Romeo and Juliet*, modeled after Hans Markart's 1882 painting.

In the Winter Garden on the main floor, visitors took in white moth orchids (*Phalaenopsis amabilis*) native to the East Indies as well as the pink Christmas orchid (*Cattleya trianae*) native to Colombia and delighted in the chocolate spots on the bearded *Laelia albida* from Mexico. Green paphiopedilums from Sumatra mingled with all manner of dendrobiums from larger Indonesia, and Central American odontoglossums and oncidiums were arrayed in viridescent banks

of club moss. Lost in a sea of tropical green and listening to the "weird music" of the band, a visitor could "almost believe himself in fairy-land."

The museum made a special

Tiffany & Company orchid brooch, New York, 1889. Likely inspired by *Odontoglossum crispum* (now *Oncidium alexandre*).

effort to court high society by displaying the most rarefied specimens. Organizers devoted the first day of each show to exclusive tours for "the most ardent admirers of this paragon of flowers;" its wealthiest patrons including Harriet Tiffany (wife of Charles Lewis Tiffany, jeweler and founder of Tiffany & Company), Ava Astor (wife of John Jacob Astor IV, author and real-estate mogul), Alice Vanderbilt (wife of Cornelius Vanderbilt II, railroad magnate), Maria Jesup (wife of Morris Jesup, banker and philanthropist), Adeline Townsend (wife of R. H. L. Townsend, silk merchant and real-estate mogul), and several wives of the extended Van Rensselaer family (historic traders, land owners, politicians, and military men). Many of these men's likenesses were added to the musée's waxworks in later years as a testament to their national importance (and perhaps no less in thanks for their wives' sponsorship of the musée in its early years). C. L. Tiffany, perhaps inspired by the orchids his wife had seen at the museum, would promote the growing love of orchids when he showed twenty-five jewel-encrusted orchid brooches at the 1889 Paris

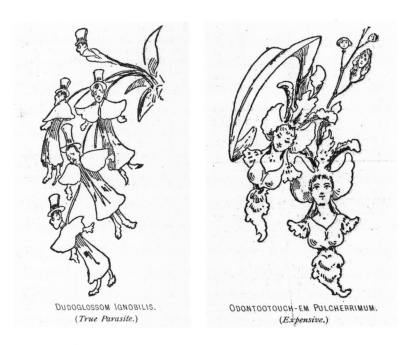

DUDOGLOSSOM IGNOBILIS.
(*True Parasite.*)

ODONTOOTOUCH-EM PULCHERRIMUM.
(*Expensive.*)

Fictitious orchids in the coverage of the 1887 Eden Musée orchid show.
The American Florist (1887).

Exposition and many more designed by Paulding Farnham in Tiffany's New York store in 1890.

As the carriages rolled up and society ladies stepped out, the media took notice. Regular visits by American aristocracy soon drew a wider fashionable crowd to the Eden Musée, as well as anyone who could muster up the price of admission. *The American Florist* went so far as to make a joke of the variety of human specimens on display: a column entitled "Orchids of the Future" included line drawings of made-up orchids. "*Odontootouch-em pulcherrimum* (Expensive)" showed busty dancing ladies on a stem and "*Dudoglossom ignobilis* (True Parasite)" revealed a dour-looking banker complete with monocle atop orchid foliage.

Along with appreciating the orchids' intricately jewel-like beauty, patrons received a lesson in Gilded Age socioeconomics. The *New York Times* assured its readers that "orchids are like diamonds, in that they have a cash convertible value at all times, always equal at least to the first cost. The difference between orchids and diamonds, however, is important. People will part with their diamonds, but with their orchids, never." The museum's rich-and-famous patrons had certainly profited in various ways from the spoils of empire in the nineteenth century—orchid extraction followed in lockstep with Western expansion.

Consider the origins of the curious and brilliant orchid gardens the musée regularly assembled: the specimens exhibited had been sourced from a veritable checklist of European colonies, foreign interests, and protectorates. Highlighted were orchids indigenous to Madagascar (where both the English and French had colonial claims through the 1880s, and France had mounted increasingly violent attacks to colonize the island nation in 1896); Central and South America (also of English and French interest); China and Myanmar (where the Opium and Anglo-Burmese Wars with the British at mid-century had assured new Western ports, sped natural resource extraction, and secured Anglo travel into the interior); India and Assam (under British colonial rule);

Opposite: A selection of orchids and tropical plants in circulation by the 1890s. "Fleurs" in *Nouveau Larousse Illustre* (1897).

FLEURS

PLANTES DE SERRES : 1. Eucharis Amazonica. — 2. Aristoloche géant. — 3. Anthurium Scherzerianum. — 4. Anthurium Laurencœanum. — 5. Bilbergia Breauteana. — 6. Bilbergia canterœ. — 7. Tillandsia umbellata. — 8. Tillandsia Lindeni. — 9. Vriesia splendens. — 10. Cinéraire. — 11. Erica. — 12. Tydœa. — 13. Eschynanthe. — 14. Gloxinie. — 15. Franciscée. — 16. Allamande. — 17. Lapagérie. — 18. Abutilon. — 19. Bougainvillée. — 20. Cyclamen. — 21. Primevère de Chine. — 22. Gardenia de Stanley. — 23. Calcéolaire. — 24. Angræcum sesquipedale. — 25. Brassavola Dygbiana. — 26. Cattleya Dowiana. — 27. Cattleya labiata. — 28. Coryanthes macrantha. — 29. Cypripedium Boxalli. — 30. Cypripedium Burfordiana. — 31. Cypripedium Io. — 32. Cypripedium Sanderianum. — 33. Cypripedium Rothschildeanum. — 34. Dendrobium phalœnopsis. — 35. Dendrobium Venus. — 36. Disa grandiflora. — 37. Epidendron vitellinum. — 38. Lælia grandis. — 39. Lælia purpurata. — 40. Lycaste Skinneri. — 41. Masdewallia Harriana. — 42. Maxillaria Sanderiana. — 43. Miltonia phalœnopsis. — 44. Miltonia spectabilis. — 45. Odontoglossum grande. — 46. Odontoglossum vexilarium. — 47. Oncidium papilio. — 48. Phalœnopsis grandiflora. — 49. Selenipedium caudatum. — 50. Stanhopea Schuttlevorthi. — 51. Vanda cœrulea. — 52. Vanda Sanderiana. — 53. Zygopetalum crinitum. — 54. Begonia rex.

Sumatra (a Dutch colony); and Malaysia, at that time a British colony, with orchids torn out of Mount Ophir and the wider archipelago's Sunda Islands in the 1880s.

The United States profited from its own imperial adventures. Americans had extracted novel-looking cacti and other plants from Mexico since before the Mexican-American War (as well as *during* the war, through botanists attached to advancing military units). On the other side of the world, in the 1890s, US imperial interest in the Philippines grew. America's growing sense of manifest destiny—and want of opportunities to get rich—culminated during the Spanish-American War, when in 1898 the United States brutally wrested away Philippine sovereignty to colonize the country. From these regions, American access to oncidiums, dendrobiums, vandas, and dozens of other orchid genera grew accordingly. Representing the refined power of Western imperialism, the range in orchid color and form displayed at the Eden Musée proved to be as extensive as the global imperial imagination writ large. Clearly, there was money to be made in international resource extraction, and orchids soon became a vital part of the circuits of Gilded Age capital.

Cupid at Work in the Winter Garden. *Eden Musée Monthly Catalogue* (1892).

Oncidium cirrhosum as *Odontoglossum cirrhosum*. Warner, Williams, and Moore, *Orchid Album* (1885).

Based on its inaugural success, and with the support of new city leadership eager to crack down on vice, the Eden Musée hosted the orchid show annually for five years—from 1887 to 1890, and again in 1892—and the number of orchids exhibited multiplied from eight hundred to more than five thousand. The museum enjoyed splashy media coverage for the annual event, and nurserymen profited by displaying unusual plants their shops were keen to sell. If orchids had long been considered expensive rich-people's toys, by the end of the century, many genera were available at lower costs from resourceful growers and the Eden Musée was quick to assure audiences that orchids were not only for the very wealthy. Over the years, orchids were exhibited at the musée as cut stems in vases, mounted on trees and wooden plaques, displayed in cachepots and slotted wooden baskets. Promoters made efforts to detail the joys of collecting and displaying orchids, emphasizing their beauty in a Victorian home as table decorations, and explaining that true and sentimental orchid enthusiasts nurtured their orchids "with as much care as a sick pet, and its day of re-blossoming is watched for with anxiety."

In addition to the lesson in political geography, the dime museum's guests—ultimately cutting a wide swath across social class—also received a scientific and artistic education. The Eden Musée's later shows presented orchids laid out in beds, often organized by country of origin, so that visitors would have a sense of the classifications of genera within the orchid family and the global regions each hailed from. Visitors were trained to inspect *Paphiopedilum insigne*, from Assam and China, "with its quaint little sac and curious mouth;" as well as *Odontoglossum cirrhosum* (from Colombia and Ecuador) "with blossoms that resemble the starfish;" "the queerly shaped *Miltonia candida* that resembles the human larynx" (from Brazil); as well as "the delicate *Cattleya mossiae*" (from Venezuela), each in its own designed space.

The later shows grew in size and scope, filling the musée's main theater and galleries with blooms. Rooms upstairs were packed with additional "rare and curious tropical plants," such as the carnivorous nepenthes and cephalotus. Present were *Calanthe vestita* from Assam, the white star orchid (assumedly *Angraecum sesquipedale*, Darwin's orchid from Madagascar), white cattleyas, *Caularthron bicornutum* from South America, the monkey throat orchid *Coryanthes macrantha* from Trinidad and tropical South America, *Oncidium alexandrae* from Colombia, and *Epidendrum stamfordianum* from Mexico. Truly, as one reporter attested, "monotony is an unknown quantity in an orchid show." Year after year, the musée's halls were crowded with orchid lovers whose favorite flowers had helped them pretend to transcend class and gender barriers for a time.

The Eden Musée organized its last orchid show in 1892. Adding to the establishment's rap sheet, just three months before the exhibition opened, the manager (not the one arrested for opening on Sundays) embezzled as much as $15,000 from the musée and was found later on the lam in Europe. With the demise of the orchid show in its original setting, the spectacle of bluebloods and ruffians rubbing elbows with orchids in the most notorious district in Gilded Age New York came to a close. The dime museum shifted its programming into moving pictures, eventually remaking itself as a full-time movie house. It closed in 1915, unable to compete with the middle-class entertainments and department stores that had moved uptown.

The flowers didn't have far to travel—Madison Square Garden, the next stop for the largest public orchid show in the United States—was only ten blocks north and picked up in 1893 where the dime museum left off. But the shows at the Eden Musée could not be duplicated, staged as they were in a nest of exotic dancers, horrifying waxworks, and notorious rowdies. (My orchids grow in similar conditions today—don't yours?)

ONCIDIUM SHARRY BABY

The grandparent oncidium species to the Sharry Baby hybrid had been systematized by 1876, so any of them could have appeared at the Eden Musée in its heyday. The ancestor it most resembles is *Oncidium leucochilum*, a Central American orchid with brown spotting and pink highlights on its petals and a prominent white lip. Or as a visitor to the Eden Musée once described oncidiums: "gaudy butterflies that flaunt themselves to the air like overdressed babies." Gaudy, sexy baby dolls—that sounds about right for the Tenderloin.

Orchid Details

~ Place of origin: grandparent species from New World tropics
~ Blooming season: usually fall–winter
~ Flowers last for: months (but you'll also wait months for spikes to develop)
~ Plant size: medium-large (can get to 3 feet tall)
~ Flower size: 0.5 inch to 2.4 inches wide; dozens to hundreds on each 2.5-foot branched inflorescence
~ Fragrance: syrupy chocolate and vanilla with plastic overtones
~ Plant habit: sympodial; large, plump pseudobulbs, leggy leaves

Orchid Needs

~ Light: medium (3,000 foot-candles), black spots possibly indicate light is too high or plant needs more moving air
~ Temperature range: intermediate; prefers winter nights above 60°F, maximum temp 85°F

- Humidity: 40%–60% ideal; can withstand lower
- Water: higher in summer, lower in winter; don't allow pseudobulbs to shrivel
- Growing medium: medium bark, perlite and charcoal mix; also does well in semi-hydroponics
- Fertilizing schedule: does best with a half-strength balanced orchid fertilizer that includes macro- and micronutrients; black leaf tips can indicate overfertilization
- Seasonal changes: Winter nights in low 60s°F for a few weeks
- Special requirements: none

Chapter 4

Frida Kahlo's Orchid

Plants grew in and through Frida Kahlo's body. They tethered her to the earth. As revolutions roiled, personal relationships gave way to heartache, and she lost baby after baby, living green plants extended to her a tenuous hold on beauty in the world. It was rare to see Kahlo without vines embroidered upon her skirts; rarer still to see her without flowers woven into her hair. In her garden at her home *Casa Azul*, in a barrio of Mexico City, Kahlo cultivated calla lilies, marigolds, sunflowers, jasmine, and prickly pear cactus flowers. She surrounded herself with bougainvillea, fuchsia, lantana, lilac, and roses.

Kahlo was an accomplished still-life painter in addition to her signature self-portraits. While she also loved dolls, toys, jewelry, exotic pets, and pre-Columbian art, which she studied, collected, and made into

Gisèle Freund, *Frida in Her Garden at Coyoacán* (1951).

Frida Kahlo, *Roots*, 1943. © 2021 Banco de México Diego Rivera Frida Kahlo Museums Trust, Mexico, D.F. / Artists Rights Society (ARS), New York. Private Collection. Photo © Sotheby's/Bridgeman Images.

art, her canvasses explode with floral abundance. From her paintings grow birds of paradise, zinnia, flowering fruit trees, irises, and dozens of other plant species. Often, the thriving vegetation in her paintings steal the show—her art was, quite pointedly, fertile.

In her personal fashion, garden, and art, Kahlo almost exclusively relied upon native Mexican flower species. Mexico is home to an extensive list of New World orchid genera, including ever-adaptable brassavolas, multicolored encyclias, spidery epidendrums, robust cattleyas, star-shaped laelias, triangular lycastes, frilly-skirted oncidiums, and dozens of others, dripping in blooms every month of the year. Her husband, Mexican muralist Diego Rivera, loved orchids, and went out of his way to find and present them to her as gifts. And, so, I've always wondered, why aren't there more orchids in Frida Kahlo's art? How is it that only a single orchid—a giant lavender cattleya—is prominent in one of her major works of art, though one that would become the touchstone of her career?

In April 1932, Kahlo was twenty-four years old when she and Rivera, twenty years her senior, moved to Detroit, where he had been given a commission from Edsel Ford (then president of Ford Motor Company and son of Henry Ford) and William Valentiner, director of the Detroit Institute of Arts, to paint a series of murals in the museum's Garden Court celebrating the "new race of the steel age." Still maturing as a painter and a wife, Kahlo worked to emphasize both her artistic and domestic selves. She searched out Mexican shops so that she could make Rivera's favorite dishes over a hotplate in their tiny apartment and dutifully delivered him lunch every day at the institute. For a time, she attempted to be the dainty, deferential wife to his rotund and towering figure.

But there were many American societal norms Kahlo did not care to adopt. During the year they spent in Michigan, Rivera and Kahlo were regularly invited to events among Detroit's moneyed elite. Kahlo soon found the small talk of upper-class ladies unbearable and was openly sickened by excessive consumption in the midst of the Great Depression. Her boxy, bright pre-Columbian inspired fashions rejected the sleek muted look of feminine American chic. At fancy events, she laced her speech with English expletives and then pretended to not know what they meant. When invited to tea at Henry Ford's sister's house, Kahlo lectured about the virtues of communism, and at a different event loudly asked Henry Ford, a virulent anti-Semite, if he were Jewish. What was more, in spite of the source of her husband's patronage, she supported the local automotive unions' rolling strikes against repeated pay cuts by their employers. She did not fit in, and she yearned for home. And then more hardship struck.

During that long, hot Detroit summer of 1932, Kahlo suffered a miscarriage at three-and-a-half-months gestation. Doctors confirmed that she had been carrying the boy she wanted—her little "Dieguito."

Two years earlier, when she was still a newlywed, physicians had ter-
minated a pregnancy early to protect her fragile health. They believed
she could not carry a baby to term due to the life-long effects of a dis-
figuring trolley crash she had suffered as a young woman—a metal rod
had impaled her abdomen, her pelvis was broken, her spine, right leg,
and collarbone were fractured, and multiple ribs smashed. She would
undergo numerous painful surgeries as a result and wore medical corsets throughout much of her life. Now, in Detroit, she had swallowed
a prescribed abortifacient soon after the first month of pregnancy, but
it did not have the intended effect. When she realized that a child was

growing within her, she imagined that a
healthy pregnancy might be possible.

After the miscarriage, as her grief
turned to boredom and then into anger,
Kahlo began to paint in earnest. She
begged doctors to let her see the fetus, as
well as to lend her obstetrical books with
illustrations that she could sketch from.
They denied her requests, but Rivera
gave her an anatomy text. The obsession
with death and decay was nothing new;
she had collected such texts at home in
Mexico and kept several skeletons and
medical treatises for reference. Her fascination with the morbid elements of the life
sciences ran deep: as a girl, she suffered
polio, and didn't shy away from representing deformation and necrosis in her art

Oncidium leucochilum is endemic from
southern Mexico to Honduras. *Paxton's
Magazine of Botany* (1840).

after her accident. She had at one point wanted to become a doctor and had delighted in human and plant biology throughout her life, examining cells and plants with a microscope. A few years after the loss in Detroit, a physician friend gave her a small fetus preserved in alcohol, which she displayed in her bedroom among her dolls. Ill health and morbidity hung over her constantly—a friend once reckoned that Frida "lived dying."

Henry Ford Hospital, painted in the weeks following her 13-day stay, was the first in a long series of garishly bloody self-portraits Kahlo created. Here, a hospital bed floats in between a dusky blue sky and plain brown earth; the only fixed object in the painting is Ford's River Rouge automotive assembly complex, looming on the horizon. A weeping Kahlo stares into the distance as she lies twisted atop blood-stained sheets, holding six objects tethered by red umbilical cords: a medical mannequin of a female reproductive system and lower spine, a male fetus, a snail (representing the slow pace of the miscarriage), a pelvis with coccyx, and a steel autoclave for sterilizing medical instruments. She knew well the autoclave from many other traumatic hospital stays. And instead of power and progress, for her the machine represented only helplessness and despair. Strikingly, she also painted a wilted lavender cattleya, resembling a collapsed uterus.

All of the tethered objects depicted in *Henry Ford Hospital* were sketched from items Kahlo observed close at hand. Rivera had brought the cut flower to her in the hospital, and the coloring and shape of the orchid in the painting reveal the same precision Kahlo brought to the medical instruments and human anatomy. All had become haunting symbols of pain, failure, and loss. Appropriately for Kahlo, the phrase for still-life in Spanish is *naturaleza muerta*, or "dead nature."

Cattleyas show how tough yet fragile orchids can be—much like Frida. The sheaths on new growth emerge slowly at first, and it is during their development into flowers that things often go awry. The sheath may be "blind" (carry no flower), the bud may be attacked by bugs, the flower may come out deformed. Cattleyas remind us of mutability—

Frida Kahlo, *Henry Ford Hospital (La cama volando)*, 1932. Painting © 2021
Banco de México Diego Rivera Frida Kahlo Museums Trust, Mexico, D.F. /
Artists Rights Society (ARS), New York.

warning us of the dangers of procreation, the unsettled waters so many
of us swim in, during our own experiences of miscarriage, pregnancy,
and childbirth. Some say that the pleasure of growing cattleyas is tied
to this tenuous moment, this first kiss foretelling an unknown out-
come. Kahlo reminds us that in gestation, plant cultivation, and the
larger human condition, we often veer from the mechanical to the sexual, from the scientific to the sentimental.

Despite the tragedy surrounding its composition, *Henry Ford Hospital* would become the grim portal to the rest of Kahlo's artistic oeuvre:
following the hospital stay, the remainder of the year that Kahlo spent
in Detroit was one of the most artistically productive periods in her

life. At first, she had sketched a bleak self-portrait in the hospital to pass the time. Soon, she bargained with herself: perhaps being a productive artist would prove more fruitful than motherhood. She then resolved to make a painting every year. In her later work, she continued to focus on the themes of her broken and bloody body, extreme loss, the mechanical versus the natural world, colonizer versus the colonized, death versus fecundity. Her paintings shocked the sensibilities of 1930s audiences with their surrealistic elements and depiction of naked pain. A fellow artist later evocatively summarized Kahlo's artistic style as offering her viewers "a ribbon around a bomb."

The cattleya orchid Kahlo painted in the following weeks of that fateful July was an attempt to record the truth of that terrible moment as well as its symbolic resonance. And interestingly, the color, shape, and relative size of the orchid in the painting all identify it as one that would have been most readily available in any large American city that year.

Much of the industrialized world was suffering through the economic crisis of the 1930s, but cattleya orchids remained a cherished luxury item for the wealthy. The cattleya corsage was an essential element in any fashionable woman's ensemble—an accompaniment that was valued for its spectacular size, scent, and beauty, and we might suspect, no less for its bald reference to human sexual anatomy—adding freshness and exoticism to the compulsory gown, hairstyle, heels, and jewelry. For the next several decades, it was the spikier, frilly-lipped, multicolored cattleya (rather than the rounded petals and circular flower shapes that distinguish most phalaenopsis, cymbidiums, and dendrobiums today) that defined the requisite floral statement.

The American orchid industry radically refocused to supply the public with cut-flower cattleyas for special events year-round. Because hybridization by cross-pollination and seed production can take seven or eight years to create new cattleya flowers, growers at the time largely relied on divisions and imports of Central and South American native cattleya species to fill the demand. By the 1940s, Thomas Young Orchids—long the largest cut-flower producer on the East Coast—advertised monthly

Woman with cattleya
corsage. *American Orchid
Society Bulletin* (1946).

in the back of the *American Orchid Society Bulletin*, pleading with private cattleya growers to sell the company their flowers so that the orchid firm could keep up with its commitments to distributors.

The wider public had far less access to orchids during the Great Depression, of course, but the flowers still held their interest. In March of 1933, thousands of Detroiters jammed the Belle Isle Conservatory, just seven miles from Frida and Diego's apartment, to see Mexican and Central and South American orchids from the collection of newspaper heiress and socialite Anna Scripps Whitcomb on display. The *Detroit News* described the orchids as both dangerous and alluring, "evocative of wild and distant scenes, of thrilling chapters in the life of some dauntless plant explorer among the forests of Brazil." "To the thrilled gaze of thousands, this week," the paper trilled, "orchids remain the symbol of glamor, of romantic loveliness," while "young men, noting the eager light in the eyes of the girl beside them, will jingle their key-rings and wonder how many luncheons they could do without to afford an orchid on their budget."

Just a few weeks before the 1933 flower show, Diego's murals made their own public debut, and the initial response was overwhelmingly negative. While Detroit's smart set would find the display of Mexican orchids "thrilling as a jungle movie," editorials condemned the art of one of the most celebrated artists of Latin America as blasphemous, "foolishly vulgar," and "un-American," demanding that the walls of the Detroit Institute of Arts be immediately whitewashed. (They weren't successful; the frescoes are still a stunning feature of the museum.) Just one month prior, the newspaper had run an article about Frida's work, condescendingly headlined "Wife of the Master Mural Painter Gleefully Dabbles in Works of Art." To the provincials in Detroit, the unusual was a thing best contained in a hothouse environment.

Given when she painted *Henry Ford Hospital*, we can deduce the particular cattleya species that Frida depicted. *Cattleya labiata*, *perciviliana*, *trianaei*, and *shroederae*, while all popular as cut flowers, do not bloom in the summer. *Cattleya mossiae*, although similarly lilac colored and long-lasting, also blooms too early,

Rossioglossum grande (formerly *Odontoglossum grande*) is endemic to southern Mexico and Central America. *Paxton's Magazine of Botany* (1841).

Frida Kahlo and Diego Rivera with a monkey in their Casa Azul garden, 1948.
Wallace Marly/Archive Photos via Getty Images.

taking center stage in the flower market in March through May. Some varieties of *Cattleya gaskelliana* look similar to Kahlo's orchid, but are lighter in color and more often filled out the wedding and graduation season in May and June. Only *Cattleya warscewiczii*, popularly called *Cattleya gigas*, reliably blooms in July.

Cattleya gigas (Latin for giant) bears the largest flower of the genus. The orchid's pseudobulbs alone can grow to sixteen inches tall, making a blooming specimen over three feet tall, and a single flower measures nearly one foot across. Other attributes underline its immense size: its flower spikes stand vertical, instead of tracing a cattleya's more typical horizontal growth, and each inflorescence can produce up to ten flowers, each bloom making a corsage. Most often, the thin oval lavender sepals and larger petals of *C. gigas* frame a slightly ruffled and deeper purple-red lip. Registered in 1854, it has been used to create hundreds of primary hybrids and tens of thousands of successful progeny. It is endemic to Colombia—one of the planet's great biodiversity hotspots—where it's called *Flor de San Juan* and *Flor de San Roque*. Unseen in Kahlo's rendering are the two yellow-white spots deep in the throat, evoking both testicles and ovaries, and perhaps reminding Frida of both her miscarriage and the loss of her son. Botanists today know that many orchids are in fact both male and female: one defining factor of what makes an orchid an orchid is the column—its sexual parts (the stamens and pistil) are fused together into one organ.

Kahlo revealed in an interview a few years later that when she painted the orchid, she "had the idea of a sexual thing mixed with the sentimental." As Frida well knew of her American counterparts, "nothing, to the feminine taste, wears like an orchid," a symbol of their femininity and class status. Yet, the orchid, given to Kahlo by Rivera—very much a *gigas* in the modern art world and with a never-ending parade of women he took as lovers—focused all of her losses into this one symbol from nature. While Diego was busy painting the Ford-inspired larger-than-life *Detroit Industry Murals*, Kahlo examined her fading cattleya, time running out on her marriage and idealistic youth.

Kahlo never again painted a sizable orchid in her major works. This was not for lack of having the flowers as a frequent presence. At their home in Mexico City, Rivera employed a local plant collector, Teódulo Chávez, to procure orchids and cacti and teach the famous artist how to cultivate them. And in 1937, when the exiled revolutionary Leon Trotsky and his wife Natalia came to live with the couple, Trotsky would appear, unannounced and in the middle of the night, demanding that he and Chávez go hunting for orchids for Frida during his months-long affair with her. Julien Levy, the owner of a small surrealist gallery in New York who mounted Kahlo's first solo show in 1938 and who was also her some-time lover, described Kahlo as a "mythical creature, not of this world—proud and absolutely sure of herself, yet terribly soft and manly as an orchid." Although we rarely think of orchids as manly today, the flowers were thought powerful and masculine as far back as the ancient Greeks; throughout time and in different cultures, they would additionally take on meanings of perseverance, courage, and virility. Orchids, like Kahlo, were alluring in their quixotic shape, their intersexual references.

Kahlo and Rivera divorced in 1939 and remarried in 1940. To mark their remarriage, Kahlo painted *The Flower Basket*, a 25-inch tondo of dozens of flowers native to Mexico. Kahlo included blooms that tradition-

Frida Kahlo, *The Flower Basket*, 1941. © 2021 Banco de México Diego Rivera Frida Kahlo Museums Trust, Mexico, D.F. / Artists Rights Society (ARS), New York. Private Collection. Photo © Christie's Images/ Bridgeman Images.

ally symbolize love and marriage, such as roses, dahlias, and jasmine, but also added zinnia, sunflowers, daisies, impatiens, calendula, hibiscus, and morning glories. Near the center sits a red-and-white striped orchid, likely *Cychnoches egertonianum*, known commonly as a swan orchid, widespread throughout Mexico and Central America. The petals and sepals of this orchid naturally curve backward, making its long, curved column more prominent, resembling a swan's neck. It is one of Kahlo's most emotionally untroubled works, but even here there are signs of pain. A hummingbird tops the tondo; the ancient Aztecs believed hummingbirds were brave little fighters, symbolizing reincarnated warriors, ready to do battle again. The bird also has meanings like love, open communication, and good luck. Yet the painting includes a fly that sits upon a flower that is past its peak. Perhaps most telling, Kahlo gave this work to American

Frida Kahlo.
*Bettmann/Bettman
via Getty Images.*

actress Paulette Goddard, the woman with whom Rivera enjoyed a public love affair, and a central cause of the artist couple's divorce. Regardless of the tondo's intent, Rivera's philandering did not cease.

The mystery of why Kahlo did not paint more orchids may be precisely because they were so over-determined as signifiers of crushed hopes, of failed love affairs, of wealthy women's artless attempts at pinning their sexuality to their chests, and of the commerce of beauty. *Cattleya gigas* recalled a moment in her life that was both full of change and full of loss. The flower was a meaningful symbol—one both masculine and feminine—at a point in her life when she likely realized that she would never bear children. Frida had tied herself to the orchid at a moment of personal crisis, and like all living things, it had wilted and collapsed in on itself. In *Cattleya gigas*, Kahlo had found a vivid metaphor for life and death, renewal and decay. She continued to create art on those themes for the rest of her life.

CATTLEYA WARSCEWICZII

Frida Kahlo's orchid, *Cattleya gigas* (now *Cattleya warsce-wiczii*), is very large and very lavender (it also has a white horticultural form). The stout orchid is a young bloomer, and enthusiasts call *gigas* one of the strongest, most persistent orchids they know. Because C. *gigas* is truly a giant, it isn't ideal in small spaces. But fear not—there are plenty of compact lavender cattleyas to choose from. Another orchid in the cattleya family that an indoor grower might cultivate instead is *Cattlianthe Jewel Box* 'Dark Waters.' It is fire-engine red, prolific, and bears a heady rose scent. *Time Magazine*, reviewing one of Frida's New York shows in 1938, wrote that her paintings "had the daintiness of miniatures, the vivid reds and yellows of Mexican tradition and the playfully bloody fancy of an unsentimental child." An orchid grown with Frida in mind could appropriately be ruby red—for her *patria*, for her firebrand art and political beliefs, for her sexual vibrancy, for proof of life and death held in constant turmoil. It's a good choice if you'd rather celebrate her sprightly side than memorialize her pain.

Orchid Details

- Place of origin: Northern Colombia
- Blooming season: late June through July
- Flowers last for: two weeks to one month
- Plant size: large, in bloom can be up to 3 feet tall
- Flower size: large, 12 inches across
- Fragrance: medicinal, hyacinth, violets, sweet baby powder
- Plant habit: sympodial, but flowers and leaves grow vertically

Orchid Needs

— Light: very high light, some direct sun
— Temperature range: nights in the 60s°F, days in the high 80s°F
— Humidity: 70% all year ideal, will cope with lower
— Water: high in late spring and throughout fall—up to 8
  inches per month, declining to 3 inches per month in winter
— Growing medium: coarse, fast-draining bark and perlite
— Fertilizing schedule: half-strength orchid fertilizer with
  micronutrients weekly when plant is in active growth
— Seasonal changes: grow slightly cooler with very low water
  in winter; if leaves or pseudobulbs shrivel, water a bit more
— Special requirements: repot only when you see new root
  tips forming

PART II

The Science of an Obsession

Chapter 5

ℱ

Rafinesque's
Strange Collections

I n 1836, continental expansion in America was picking up speed, and scientific and philosophical inquiry were keeping apace. In that year, Wisconsin Territory was formed, the Alamo fought, and the Republic of Texas established. In late December, a sudden freeze with 70 mph winds made temperatures in Illinois drop 40°F in a matter of minutes. Travelers, animals, and anyone who could not find shelter in time suffered extensive frostbite or death. Western states and territories held surprises, not all of them good.

That same year, Constantine Samuel Rafinesque, an anxious polymath living in Philadelphia, wrote a book—one of six that he published in a matter of months—with a warning for would-be botanists. Anyone who wished to be a "pioneer of science" must be fully prepared to meet dangers of all sorts in America's wild groves and mountains. Readers opening his *New Flora and Botany of North America*—hoping to learn about wild and unknown flowers—were regaled with the naturalist's nightmares. He recounted the "gloom of solitary forests . . . when the fare is not only scanty but sometimes worse; when you must live on corn bread and salt pork, be burnt and steamed by a hot sun at noon, or drenched by rain."

Travel and weather conditions were unpredictable at best when on the hunt for native North American orchids. Rafinesque had made a vigorous, if not profitable, career searching for them in the wilds of Kentucky and while touring several other states. He attested to the multiple hardships of traversing the land, among them how naturalists met "rough or muddy roads to vex you, and blind paths to perplex you, rocks, mountains, and steep ascents." Having no respect for those who chose an easier course, he railed on sedentary botanists, and those who

Portrait and frontispiece to C. S. Rafinesque, *Analyse de la Nature* (1815).

traveled in carriages and by steamboat. Naturalists who herborized merely near their homes did not face danger, and thus did not deserve to find botanical wealth. Rafinesque was one of dozens of professional plant hunters combing the quickly incorporating US countryside for unnamed species. Born in Constantinople and raised in France and Italy to French and German parents, he first came to America in 1802. After a stint in Italy as a businessman, he returned to the United States in 1815, where he stayed for the rest of his life.

The story of North American botanical exploration has many epi-centers. Centuries prior to European contact, Native people in hundreds of tribes had exquisite understanding of the plants that surrounded them. So, Western knowledge of American plants is not as much a tale of "discovery" as documenting many plants that were already known and putting them into Linnaean nomenclature, official botanical cat-egories, and publishing them. By the early 1800s, institutions like the Royal Horticultural Society and the American Philosophical Society deemed some types of information valid and others invalid by virtue of who published the information (white men were favored), where (west-ern Europe and America), what type of education they had (from elite universities), and whether it conformed to newly consolidated fields of Western science. Scientific collections were limited to those people and plants that were able to make it into the logs via accepted institutions.

Yet even within these societies crept a few wild cards. Though he is little known outside of botanical circles, the well-traveled pansophic C. S. Rafinesque (1783–1840) was one of the most prolific writers of the nineteenth century. Many also believed that he was insane. As a fellow transplanted North American botanist attested, "he is doubtless a man of immense knowledge—as badly digested as may be & crack-brained I am sure."

Rafinesque spent weeks with John James Audubon in Kentucky in 1818. One night the famous ornithologist witnessed a stark-naked Rafin-esque, broken violin in hand, swinging it wildly at bats that had entered his room through the window. Rafinesque, once exhausted, begged

Audubon to help him capture the creatures, sure that they were a new species in need of classification. Audubon reluctantly complied, administering a "smart tap to each of the bats as it came up, and soon had specimens enough." We might read this scene in two ways: Rafinesque was a naturalist willing to upset a household for potential discovery— or he was a delusional man driven to attach his name to anything that moved. Rafinesque was the same with plants. When Audubon showed Rafinesque what might be an undescribed plant species on a riverbank, he "thought [Rafinesque] had gone mad. He plucked the plants one after another, danced, hugged me in his arms, and exultingly told me that he had got not merely a new species, but a new genus."

Rafinesque was excited because his life's passion was scientifically reclassifying plants and animals. "Species" are the basic unit of biological classification; "genus" is one step above that, collecting similar species within its rank. The higher up his reclassifications climbed, the more significant the change to science. Genera (the plural of genus) link groups of orchids that have similar traits (and today, similar DNA). For example, when Rafinesque attempted to rename the South American orchid *Oncidium papilio* to *Psychopsis picta*, he was suggesting both a generic and species name change, but left it in the family orchidaceae (family is the major rank above genus). He was successful with "psychopsis" because he was able to prove that the genus held different traits than oncidiums, but the "papilio" remained unchanged because the original species name properly followed Linnaean naming conventions. The size of each orchid genus is widely variable; for instance, the genus bulbophyllum contains two thousand species, while mexipedium is a genus with only one species.

Whether Rafinesque was mentally ill or not, he made significant contributions to biology, linguistics, and anthropology. He gave Latin names to more than 6,700 plants—well over 100 of them orchids—and attempted to describe sixteen types of lightning for science. Like any good botanist, Rafinesque lovingly tracked the seasonal phenology of local plants. In

ONCIDIUM PAPILIO MAJUS.

Rafinesque renamed *Oncidium papilio*
as *Psychopsis papilio*. Warner, Williams,
and Moore, *Orchid Album* (1887).

Rafinesque renamed *Orchis spectabilis* as *Galearis spectabilis*. Robert Sweet, *The British Flower Garden* (1823–1829).

early May 1816, outside of Philadelphia, above the falls of the Schuylkill, he spotted the first Showy Orchid of the season (then known as *Orchis spectabilis*), among wild violets and azaleas. He spent enough time with the orchid to realize it deserved reclassification and went on to rename this orchid of the eastern United States *Galearis spectabilis* in 1833.

By the early 1820s, Rafinesque was also adept at describing the ecosystems in which particular plants grew. He found the Kentucky hilly region, where the waters of the Kentucky, Green, Cumberland, and other rivers divided, was a mixed ecosystem of pine savannas and acidic bogs. It had proven a good place for *Orchis ciliaris* (yellow fringed orchid, now *Platanthera ciliaris*), red cedar, wild currants, pitch pines, and several other of his favorite plants to grow.

Perhaps the North American orchid he was most fond of was the yellow lady's slipper, *Cypripedium parviflorum*, which he attempted (but failed) to rename *Cypripedium luteum*. He wrote about this orchid at length and included it in his 1828 *Medical Flora of the United States*. Here Rafinesque followed the tradition, thousands of years in the making, of identifying plants for their medicinal properties. He then appended their scientific interest, possible market value, and any other knowledge he had gained (and hoped to be made famous for) in studying them. He detailed that the yellow lady's slipper bloomed in May and June from New England to Louisiana, and was a valued flower in regional gardens. He added that it had long been known to Native people, who called it Moccasin Flower; they used it medicinally, as a hair ornament, and to induce an abortion. He promised cypripediums alleviated all nervous diseases and "hysterical affections" by allaying pain, quieting the nerves, and promoting sleep. (And indeed, lady's slippers are used in various homeopathic preparations as well as sedatives in supplements today.)

Rafinesque collected plants across fourteen states, covering more than eight thousand miles. One imagines him strutting across the continent, boisterous, pointing and shouting "all new!" at plants, clouds, animals, and people, as he used the phrase repeatedly in his journals and publications. He was also employed for seven years as a professor of botany at Transylvania University in Lexington, Kentucky, but was fired for quarreling with the university president and faculty. Rafinesque soon become a terrible bane to established botanists like Asa Gray at Harvard. He sent editors of several respected botanical journals so many new species and attempted to divide so many genera—many of them poorly described or described only from other naturalists' notes—that editorial staffs gave up reviewing his manuscripts, or banned his work from submission all together. Rafinesque then turned to self-publication

No. 30.
CYPRIPEDIUM LUTEUM.

YELLOW LADIES' SLIPPER.

Cypripedium parviflorum as *Cypripedium luteum* in Rafinesque, Medical Flora, or, Manual of the Medical Botany of the United States (1828).

Rafinesque with plants, c. 1810.
Possibly painted by William Birch or
Falopi.

and other ventures, attempting and repeatedly failing to establish several
of his own horticultural magazines and a botanical garden.

As a botanist, he attested that he met "all kinds of adventures, fares
and treatment. I have been welcomed under the hospitable roof of
friends of knowledge or enterprise, else laughed at as a mad Botanist
by scornful ignorance." Yet he also admitted, taking on the voice of a
jubilant war veteran, that he was healthier and happier on the trail—
the pleasures of exploration outweighed its miseries and dangers. He
described sound sleeps and soothing naps "under a shaded tree near a
purling brook" as some of his great delights while herborizing.

Writing of his collecting habits and plants-turned-friends, "here is
an old acquaintance seen again; there a novelty, a rare plant, perhaps
a new one! greets your view." Rafinesque describes hastening to pluck
new specimens, examine them, and put them in his next book. Walking on, he felt exultation in making a "conquest over Nature." There
was of course power in adding a page to science, giving names to living
things, transforming specimens into knowable and useful objects. And
when bored on the trail, he felt he communed with a higher power. In
his acquaintance, "Every pure botanist is a good man, a happy man,
and a religious man."

In the age-old battle of taxonomic lumpers versus splitters, Rafinesque was a consummate splitter. His *Flora Telluriana* (meaning "flowers of the earth" and published in 1836), the most detailed of his botanical works, was an attempt to establish two thousand new plant genera. Dozens among these, he demanded, were to be split away from established orchid genera. In a matter of about six pages, he renames orchid species from Jamaica, the United States, India, Nepal, and Mexico. With typical hyperbole, he compared the necessity of splitting orchid genera to how another family of plants needed to be sundered: "whoever unites *Azalea* to *Rhododendron* sins against Linnaeus and Nature!" The comment is funny, but puzzling. As a proponent of the natural system, Rafinesque often attacked Linnaeus's sexual system to find footing for new plant names. On the other hand, the erratic botanist wanted nothing more than to step into Linnaeus's shoes as the leading plant systematist for a new century. And the gardeners among us will realize that Rafinesque often did not win out over the lumpers in his time: still today, azaleas are within the genus rhododendron.

Rafinesque went so far as to openly attack the most famous botanists of his day, including Sir William Jackson Hooker at the Royal Botanic Gardens, Kew, and John Lindley, orchidologist and chair of botany at the University College of London, who he thought bundled too many plants into a single genus. Rafinesque's peers in America stood up in defense of all botanists by attacking him right back: one wrote that Rafinesque's greatest "misfortune" was his "prurient desire for novelties and his rashness in publishing them." Another marshaled evidence to prove that Rafinesque had invented species out of "his love of fame and insatiable species-mongry."

In terms of new orchid genera, Rafinesque attempted to add more than one hundred of them to botany. One of his chief concerns was splitting up the genus epidendrum. A name at one time given to almost any orchid that grew on trees, Rafinesque was accurate in calling it a "confused medley" of plants. In *Flora Telluriana*'s first entry, he proposed dividing night-scented orchids from epidendrum and calling them *Nyc-*

Posthumously, Rafinesque was successful
in renaming the genus psychillis. *Psychilis
atropurpurea* is here *Epidendrum atropurpureum*
in *Lindenia: Iconographie des Orchidés* (1886).

tosma ("sweet smelling by night"). One of Rafinesque's chief blunders throughout this work and others, however, was to attempt to rename genera and species without having the orchids in front of him. His attempts at the generic re-definition of eight dozen orchids throughout his *Flora Telluriana* astoundingly included plants from North and South America, throughout the Caribbean, Africa, Portugal, India, Nepal, and Russia. This was a problem because no one could be expert enough to reclassify one hundred orchids from such diverse regions, let alone reorganize them within Linneaus's sexual system without the orchid flower in front of them. He also attempted to reanimate generic names he felt had been overlooked, as well as ask Western botanists to consider the work being done by local people in various orchid-rich countries. Far ahead of his time, he felt "the Japanese, Chinese, Hindu, and Arabic botanists deserve commemoration like ours" in the annals of scientific botany.

If we gauge Rafinesque's success rate with orchids by the scientific names that stand today, roughly more than 10% are still in use (which isn't bad, given the changes DNA analysis has wrought over the past few years alone). But most orchid genera Rafinesque penned don't roll off the tongue easily. Say some with me now: *Caularthron* (meaning

Rafinesque renamed the genus tolumnia. *Tolumnia guttata* as *Epidendrum guttatum/Cymbidium guttatum* in *Dictionnaire des Sciences Naturelles* (1816–1829).

"stem jointed," from tropical America), *Eltroplectris* ("free spur" from south Florida and the Caribbean), *Sacoila* (meaning "bag hollow," from Florida and Mexico), *Plectrelminthus* (the "spur worm" orchid genus from Africa), and coralroot *Hexalectris*, among others found in the eastern United States, Mexico, and Central America.

Rafinesque-named genera more popular with indoor growers today include psychopsis, with its ornate purple- and green-spotted leaves. Rafinesque wrote that the new genus should be separated from oncidium because the pseudobulbs, leaves, and flowers were distinct, with large variegated yellow and orange flowers. With kaleidoscopic maroon and green leaves, purple pseudobulbs, and canary yellow and pumpkin orange blossoms, it is one of the orchids that triggered Victorian orchidomania and is still beloved for its astonishingly large sequential blooms and very tall inflorescences.

Rafinesque also found success with the petite and frilly genus tolumnia (from Central America and the Caribbean). Although he renamed it in 1837, in the orchid trade, dancing lady tolumnia orchids have only been hybridized for the past fifty years, with interest peaking in the dainty, big-lipped orchid in the past decade. They grow best mounted or as air plants, with high humidity and constant air movement. Orchid growers also still appreciate Rafinesque's cochleanthes, the "shell flower" of American tropics, and psychilis, the "peacock orchid" from Hispaniola.

Unsurprisingly, in the long term, Rafinesque had his best luck systematizing orchid species to which he had access. He successfully named a species of the North American summer coralroot (*Corallorhiza maculata*), the showy orchid ("helmet like" *Galearis spectabilis*), two small, green-flowered bog-loving pogonias (in the genus *Isotria*), the East Coast fringed orchid (*Plantathera x bicolor*), the southern twayblade (*Neottia bifolica*), as well as slender ladies' tresses and little ladies' tresses (*Spiranthes lacera* and *tuberosa*).

Amidst his endless lists of Latin and Greek names, Rafinesque made time for a more poetic appreciation of plants. In a lecture he delivered

Rafinesque encountered *Pogonia ophioglossoides*, common name Rose Pogonia, which ranges across the eastern half of North America.

in Kentucky in 1820, he described nature as a "beautiful and modest woman, concealed under many Veils, some of which she throws aside occasionally or allow[s] them to be removed by those who deserve such a high favor." And in a nod to one of his favorite naturalist-philosophers, he mimicked Erasmus Darwin in his romantic love for flowers:

> To the sun Their leafy limbs eqpand [sic],
> And nuptial buds with dazzling beauties bloom
> Of thousand shapes and hues, or sweet perfumes,
> The Earth adorning with a verdant dress,
> Sprinkled with floral gems like lucid stars.

Rafinesque died painfully of stomach cancer at the age of 57, likely provoked from extended self-medication with maidenhair fern. Once in a great while, historical figures pierce their own oddities and failings and offer us a perfect rendering of themselves. In the final years of his life, in the last installment of *Flora Telluriana*, Rafinesque wrote "I hope to become the Nestor of Botany." In Greek mythology, old King Nestor was known for his wisdom, eloquence, and bravery. Throughout *The Iliad*, he advises warriors in battle, but his coaching is often ineffective in the moment. In some depictions of Nestor, he is a snarky old coot, boasting of past accomplishments, while drowning others in misbegotten information. Wise but often misbegotten: Rafinesque encapsulated his strengths and foibles perfectly.

PSYCHOPSIS PAPILIO

Orchid enthusiasts have a plethora of species to grow in remembrance of Rafinesque. For sheer historical interest and its odd form, psychopsis will do nicely as metonymy for him. *Psyche* means soul or personality in Greek, and *papilio* means butterfly in Latin, the combined scientific name an accurate description of both the orchid and the flighty botanist. *Psychopsis* Mendenhall, pictured here, is a hybrid with *Pyp. papilio* as 75% of its parentage. If you are a week-end naturalist living in North America, you might also make a spring or summer day of hunting for native orchids named by the botanist, as well as try your hand at growing native lady's slippers in your garden. But only procure native slippers from a reputable source. Collecting them from the wild is a federal offense.

Orchid Details

~ Place of origin: Northern South America; Costa Rica to Peru
~ Blooming season: all year, peaks May to September; do not remove an inflorescence until it is dead and brown because the orchid will rebloom on a seemingly inactive spike
~ Flowers last for: months to years (flowers sequentially on same stem)
~ Plant size: when not in bloom, 12 inches tall; when in bloom, the inflorescence (flower spike) can reach 5 feet
~ Flower size: 6 inches
~ Fragrance: none
~ Plant habit: sympodial; sprawling purple and green spotted leaves and pseudobulbs (can be trained to be upright)

Orchid Needs

~ Light: moderate, no direct sun
~ Temperature range: warm, 57°F–84°F
~ Water: heavy in the summer, up to 8 inches per month; light in the winter, as low as less than 1 inch per month
~ Growing medium: coarse, well-drained bark and perlite; media must remain fresh, but the orchid resents root disturbance; repot only when new root tips emerge and only when necessary
~ Fertilizing schedule: weakly (one-quarter or one-half strength), with balanced orchid fertilizer weekly during active growth
~ Seasonal changes: very high water May through September; very low water December through April; roots should always dry out between watering

Chapter 6

✤

The Wind Orchid

L ate-twentieth-century New York orchid grower Joe Kunisch
once claimed, "You can get off alcohol, drugs, women, food, and
cars, but once you're hooked on orchids, you're finished. You
can never get off orchids . . . never." It's still true that you don't want to
get in between orchid lovers and their collections. But if you want to
meet someone with an all-out obsession, find a lover of *Vanda (Neofinetia) falcata*. For a dose of their orchid mania, try a visit to a neofinetia club, check out the prices of recently ranked Japanese *fuukiran*
varieties, or scope out the intricate beauty of the miniature cloisonné
orchid pots of the masters. You won't be disappointed.

Although the popular story of *fuukiran* in circulation today relies
heavily on Samurai tropes and other tales of Japan's historical rich-

Neofinetia falcata exhibiting its flower spurs and tsuke.

and-famous, we might also tell a story of neofinetia that embeds the treasured orchid in circuits of science and the exportation of Japan's art and culture. Threading together *fuukiran*'s multiple storylines unlocks a long history of scientific learning and mutual cultural appreciation that have been hallmarks of the East–West relationship for centuries.

Vanda falcata's story is somehow simultaneously vaunted, piece-meal, and different from all other orchid stories. Even its science is set apart from other orchids: *V. falcata* is a single orchid species that has modified itself into more than 2,200 distinct varieties. Its more popular trade name is *Neofinetia falcata*—most growers steadfastly refuse to give up the moniker made popular in the 1920s, and still lovingly call these orchids "neos" for short. These orchids are small—most varieties measure less than six inches across—and the exquisite details of each cultivar make them utterly collectible. But if we consider the country that made neos culturally important, and still sets the standard for varietal appreciation worldwide—Japan—we should instead call the orchid *fuuran*, the "wind orchid." Exceptional varieties of *fuuran*,

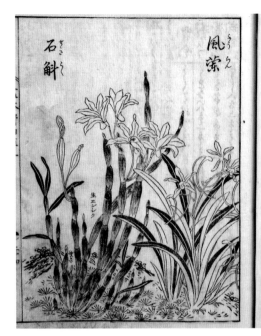

Tachibana Yasukuni, a collection of orchids (*Dendrobium moniliforme*, *Vanda falcata*, and *Phalaenopsis japonica*) in *Ehon Noyamagusa* (1755).

judged annually in Japan, are deemed *fuukiran*, "orchids of the rich and noble."

Native to Japan, China, and Korea, *Vanda falcata* has had many names over the centuries—fourteen of them assigned within the Linnaean binomial system alone, dating back to 1784. But *fuuran* was in use more than a century before that. It is possible that wind orchid appreciation in Japan dates as early as the late 1500s, when the second Tokugawa shogun Hidetada was said to harbor a "flower addiction." And by the 1700s, Japanese *fuuran* were enjoyed across East Asia for their fragrance as well as their form. This was in part due to Japan's middle Edo period giving rise to *kiju-iso*, a formalized appreciation of strange trees and unusual plants. *Fuuran*, existing in so many delightful, variegated forms, were primarily enjoyed by the daimyo—Japan's feudal lords—and other wealthy people, but less expensive and less rare varieties were collected, traded, and loved by common people as well.

Also known as the "Samurai orchid," *fuuran* holds a celebrated place in Japanese history. For more than two centuries, the Tokugawa shogunate (1615–1867) brought relative prosperity and peace to Japan with new systems of education, law, and commerce—centuries when fine arts and learning flourished. Science was developing as well. As early as 1736, Tokugawa Yoshimune organized a nationwide survey of natural species. By the mid-eighteenth century, *fuuran* were classed with other orchids and considered important and highly aesthetic garden plants in Japan. Horticulturist Matsuoka Gentatsu published a monograph on orchid cultivation early in the century; by 1755, artist Tachibana Yasukuni published *Ehon Noyamagusa* (*Mountain Grass Picture Book*), depicting *fuuran* in a scene with other treasured orchids. As historian Frederico Marcon has found, Tachibana was part of a large and diffuse set of practitioners of *honzogaku*—Japanese academics, gardeners, artists, and dealers interested in the empirical and aesthetic study of plants and animals that had begun to branch off from the more medically focused nature study of centuries earlier. If this sounds a lot

Women buying plants from street vendor in Gazu, *Shiji no yukikai* (1798).

Botanical identification and matching games had been popular in Japan since the eighteenth century. Here, girls play a flower card game in King, *Farmers of Forty Centuries* (1911).

like how the Enlightenment developed in the West, it is—except that *honzogaku* was borne of Eastern thought and only much later evolved and adapted in tandem with Western scientific thought.

Natural history in the form of collecting flowers, plants, shells, insects, and birds became fashionable for people of all classes. Gardening, botanical identification and matching games, herb-hunting picnics, and hybridizing new breeds of plants and fish were favored pastimes. Collecting wind orchids was part of this rich culture of nature study and appreciation.

One possibly apocryphal story tells that Japan's most famous shogun, Tokugawa Ienari (1773–1841; held office 1787–1837), was infatuated with *fuuran*'s crystalline white blooms—he believed the flower's heady nocturnal jasmine scent aided in the seduction of the few dozen concubines he kept throughout his reign. Samurai, cultured men, and merchants sought out new and rare *fuukiran* for him, and Ienari may have

Frontispiece to *Voyage de
C. P. Thunberg au Japon* (1796).

accepted gifts of the small orchid from his feudal lords in exchange for
large estates.

Japan was one of the most notoriously inaccessible places on Earth
for more than a century. The shogunate suffered doing business with
only the Chinese and Dutch. All trade was limited to the tiny man-
made island of Dejima in Nagasaki Bay, well over one thousand miles
from the city of Edo (now Tokyo) by boat and at the far western end
of the main island. Edo, at one million people strong in the 1770s and
one of the largest cities on Earth, was an alluring, fascinating place for
those lucky enough to visit it.

One of the early naturalists to visit Japan was Engelbert Kaempfer,
a German serving as a medical officer for the Dutch East India Company in Japan in the early 1690s. Linnaeus later made use of Kaempfer's publications on Japan, and was introduced to the country's orchids

through him, including *Dendrobium moniliforme*, an orchid that has enjoyed centuries of popularity in Japan and also around the world.

More than eighty years after Kaempfer's visit, Carl Peter Thunberg (1743–1828)—a Swede serving as a medical officer of the Dutch East India Company—arrived in Japan. Thunberg had spent three years at the Dutch Cape Colony in South Africa, perfecting his accent, so that he could pass as Dutch and would not risk punishment or extradition from Japan. Thunberg was eager to make his own mark in global botany. In less than fifteen months, he collected more than one thousand species for Linnaean science. Linnaeus himself had been Thunberg's mentor in the 1760s—the father of taxonomy once described his student as "diligent, intelligent, and unassuming."

In 1775, Thunberg's ship was one of two Dutch traders allowed into port; he was one of only two hundred Europeans allowed to set foot

Kawahara Keiga, *Arrival of Dutch Ship to Dejima*, c. 1820s.

upon Dejima that year. From that two-acre island, Thunberg successfully instructed Japanese students in medicine and botany. In return,
Thunberg's students gave him plants in thanks and unregistered payment. Some of the plants the students collected themselves, and some
were gained in trade or through friends in the interior of the country.
Thunberg was delighted—he raved that "they brought to me on the
island various plants of this country's produce, which were not only
beautiful and scarce but likewise hitherto totally unknown."

Thunberg also struck up professional friendships with Japanese
physicians Nakagawa Jun'an and Katsuragawa Kuniakira. The men
began visiting Thunberg daily, developing a rich intellectual exchange
that can only be guessed at in Thunberg's later work on Japan, as the
men kept no other records of their discussions. After many months,
Thunberg, through his successful medical practice and well-connected
friends, gained the Nagasaki governor's trust and was ultimately
allowed to "herborise" and make a few botanical excursions into the
hills around the port.

The freedom to search for plants came at a cost, however—the botanist was surrounded by guards, officers, and interpreters at all times,
comprising upwards of twenty or thirty people on each excursion.
Thunberg was also required to pay for transporting and feeding the
party. Historians now believe that the cost and effort to mount these
botanizing expeditions, along with the limited number of months hospitable enough for collecting, likely means that most of Thunberg's
collection were presents for him. His mentees brought him plants that
were highly prized, aesthetically presented, and easily carried—*fuuran*
fit the bill perfectly.

Thunberg is sometimes better known for bringing knowledge of
Japanese society and political systems back to Europe, but his singular priority was Japanese botany. As case in point, he was once given
the rare privilege of meeting an abbot of a central shogunal temple in
Nagasaki. He dismissed the experience: "It afforded me less pleasure
than the shrubs I met with in the vicinity of his church." Yet upon his

Vanda falcata, here called
Limodorum falcatum. Thunberg,
Icones Plantarum Japonicum
(1794).

LIMODORUM Falcatum.

return to Europe, the botanist was not a snob—he did not subscribe
to most of his contemporaries' beliefs that Western society was supe-
rior in all ways to that of the East. Thunberg extolled the virtues of
Japanese "Enlightenment and culture"—its *honzogaku*—the rest of his
days, but often added that he hoped that he had brought European science's "brighter rays" to Japan as well. In one respect, he was correct:
through his medical practice, he spread information about Western
treatments for syphilis—Linnaeus had taught him "cures" that were
"sure, but dangerous" involving mercury and arsenic. Thunberg taught
the method (actually a palliative, not a cure) at all locations he bota-
nized, including rare visits by special invitation to Nagasaki and Edo,
and therein remained a household name in Japan for a generation.
Yet Thunberg's scientific legacy in the East foundered, likely because
the Japanese didn't need another scientific system through which to
understand their plants. They had their own.

While Thunberg did not bring back to Europe live plants or seeds—
the living specimens he loaded on the four-vessel flotilla home were
destroyed by a storm in the English Channel, devastatingly, almost
upon his doorstep—he did save an impressive collection of pressed
specimens. His friendship with Nakagawa Jun'an continued to pay div-
idends as well—the Japanese physician-botanist sent the Swede plants

for many years, contributing yet more species to Thunberg's publication of *Flora Japonica* in 1784. The orchids and other flowers Thunberg detailed still have a place on gardeners' wish lists: iris, lobelia, weigela, viburnum, several lilies, begonia, magnolia, clematis, and ranunculus were all introduced to the West by Thunberg.

Thunberg was still extolling the most promising of Japanese orchids in a speech to the Linnaean Society of London in 1793. Recited in Latin, he waxed poetic in his prefatory remarks, "Every year there is increase of knowledge, amended every day, and little by little it culminates, with the desire of several men who work in haste. They uncover a new earth, and new moons of nature. These treasures of nature are then made known, and useful to the human race." Upon his return to Europe, his work on Japanese flora was continuously scrutinized by what he called numerous "indefatigable scientists and brisk tourists." He happily regaled his audience at the Linnaean Society with his knowledge of Japanese flora, and, as one might guess, the first plant family he highlighted for the society was *orchidaceae*, including six orchids in wide circulation still today: petite *Dendrobium moniliforme*, the long life orchid; *Cymbidium ensifolium*, the four-season orchid;

Bletilla striata as *Limodorum striatum. Curtis's Botanical Magazine* (1812).

Pecteilis radiata, the white egret flower; *Bletilla striata*, Japan's common twayblade; and *Vanda falcata*.

Thunberg first registered *fuuran* as *Orchis falcata* in 1784. Though its genus has been reclassified many times, the *falcata* stuck. Falcata in Latin means scythe, sword-like or sickle-shaped; we might surmise that Thunberg registered its name this way as both a way of describing the shape of a typical neofinetia leaf and flower spur and as a nod to the Tokugawa shogunate culture of the day. Here, in one small orchid's name, Thunberg demonstrated his much larger intentions for the knowledge he gained in Japan. His botanical treatises and cultural descriptions served as the foremost sources on Japan during Europe's late Enlightenment period—he married horticultural knowledge with political detail for Western audiences in ways that were not usually present in the scientific record. Due to Thunberg, to Western eyes, Japan was a wonderland of plants with a storied cultural history from its earliest iteration.

In addition to Thunberg's books, live *fuuran* had begun to circulate in Europe by the early 1800s. The first known live *V. falcata* in England was cultivated by Sir Abraham Hume at Wormleybury; he had received the plant from William Roxburgh, a Scottish botanist who had left Asia

Vanda falcata,
here "Sickle-leaved
Limodorum." *The
Botanical Register*
(1818).

in 1813, loaded with specimens. At the same time, *fuukiran* culture in Japan was reaching its peak. Albums of prize *fuukiran* were published, lists of named cultivars exploded, and charts of ranked specimens circulated. Although grown outside all year—Japanese homes were traditionally dark, and *fuuran* need bright light, temperature variation, and air movement around their roots to do well—when displayed the orchids were set in elaborate decorative pots. Gold and silver wire cages were placed atop the precious plants, in an effort to deter small rodents from snacking on them and to protect them during travel. When aficionados would scrutinize the orchids up close, they were expected to place a piece of paper in or across their mouths to avoid breathing on the plant—exactly the practice one would follow if inspecting a shogun's heirloom sword.

It is from this period—the early nineteenth century—that most current-day *fuukiran* cultural appreciation still stems. Unlike most orchids, an entire web of Japanese cultural expression around *fuukiran* has carried with *Neofinetia falcata* into orchid growers' homes around the world. Most collectors will tell you that they enjoy an appreciation of Japanese history, culture, and aesthetics that come with any time spent with neos. They invest time in learning the deep history of the orchid, techniques in how to set the miniature vanda in a traditional moss mound, and employ an ornamental pot to display the orchid in the Japanese style. There is simply no other orchid in circulation that carries with it so many cultural components. Its only likeness, really, is to another esteemed Japanese horticultural export: bonsai.

By the mid-1860s, images, cultural details, and a growing number of *fuuran* were in circulation across wider Europe. To the German ear, its Japanese name was "Fu-Rang," the French heard "Fou ran," and the English gleaned "Fuu-lan." The German periodical *Gartenflora* published an in-depth article about the orchid in 1866, detailing how its owner achieved regular blooms, highlighting its need for a cool rest in winter. It was a "strange, peculiar species commenda-

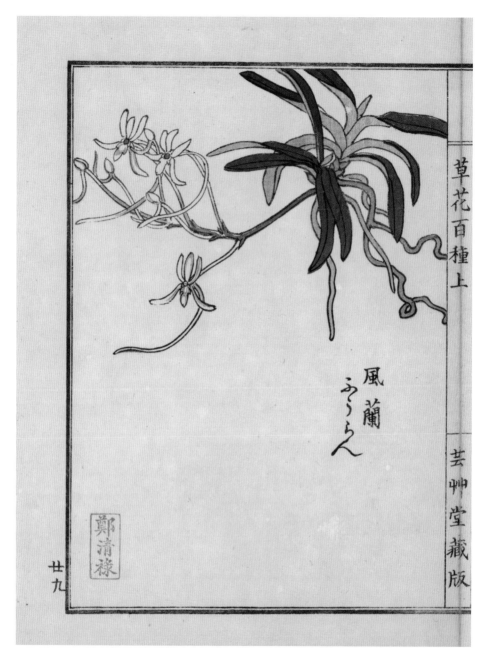

Kono Bairei, *Fuuran* from *Kusa Bana Hyakushu* (*One Hundred Varieties of Flowers*) published posthumously in 1901.

ble for the fragrance of its white flowers." The editors stated with certainty that the orchid was one of the most valuable introductions from Japan to date. One erroneous detail about the orchid was still in circulation, however—Thunberg, almost a century earlier, described that he found *V. falcata* in the mountains near Nagasaki, growing on the ground between bushes. While he had the general location correct, it took a long time for Europeans to realize that the neofinetia specimens they found on the ground had been blown out of trees by the wind.

During ensuing decades and through the turn of the twentieth century, *fuuran* enjoyed another peak of aesthetic appreciation in Japan. Kono Bairei (1844–1895) was a leading *ukiyo-e* and *kacho-ga* (bird and flower picture) painter and teacher in the mid-Meiji period, specifically within the famed Maruyama-Shijo school based in Kyoto. The school developed a synthesis of Western and Eastern styles—that is, Western realism created through traditional Japanese painting techniques. Herein, painting directly from nature was paramount. In Kono's work, the orchids are detailed, yet also project an expressiveness, a spirit beyond strict representational botanical art. Other botanist-artists such as Yohio Tanaka, Ryokichi Yatabe, and Tomitaro Makino created lush orchid art as well.

By the 1910s, the West was deep in the throes of Japonisme—a growing appreciation for not just Eastern graphic aesthetics, but also ceramics and other arts. Cloisonné enamelware was at its apex as well. Displayed at world's fairs, haute galleries, and increasingly within department stores, the Japanese aesthetic raged in Europe and America for more than thirty years. A wave of Japanese gardens and horticultural techniques soon joined the trend. Positioned directly at the center of this matrix of Japanese visual art, gardening, and ceramics were *fuukiran*. And by 1920, the wind orchid enjoyed yet another resurgence in Japan, when the All-Japan Fūkiran Society was formed. To this day, the society annually ranks the most precious *fuukiran* and generates a chart to display their hierarchy—designed in the style used to rank sumo wrestlers.

Jason Fischer, neofinetia expert and manager of Orchids Limited in Minnesota, notes that form takes precedence over flower in these small orchids. Most *fuukiran* are judged whether or not they are blooming at the time, and white-flowered species are almost always judged while *not* flowering. Unlike much of the West, in Japan, the orchid's shape, root tips, variegation, connection point between the axis and leaves (its *tsuke*), as well as presentation of the plant upon its moss mound, are all more important in defining a valuable variety than its blooms. Within these details, root tips are especially important because they come in many surprising colors: red-brown, yellow, shades of green, and ruby/hot pink. And if you're not careful, you may miss the intricate beauty of a neo's *tsuke*—the Japanese describe them as straight, round, or shaped like mountains or waves.

Unlike most other orchids, with *fuukiran* we are asked to consider each variety—and indeed individual plants—as their own unique works of art. Fischer describes that in Japan *Neofinetia falcata* is thought to be the "perfect orchid" because an enthusiast can pick it up with one hand, and with a simple turn of the wrist, inspect it on all sides. Tachibana Yasukuni, Kono Bairei, and other Japanese artists' work, along with the increasing international mobility of the orchids they painted, are a window into Japan at a time of great change. *Neofinetia falcata* had once symbolized the power and prestige of Japan's centuries-long shogunate; by the twentieth century it was a symbol of Japan's rarefied place on the world stage. Orchid and botanical art helped propel a Western rage for Japanese-inspired art—European impressionists including Manet, Monet, van Gogh, and Degas, as well as American artists and craftspeople Mary Cassatt, Louis Comfort Tiffany, and Frank Lloyd Wright, were heavily influenced by Japanese flower prints. Not limited to the late nineteenth and early twentieth century, Japonisme worldwide lives on today, as do a growing number of exquisite varieties of *fuukiran*.

VANDA FALCATA

The appreciation of *fuuran* and other miniature orchids is related to the Japanese love of bonsai and makes them highly collectible for growers with limited space, as they grow well on a cool, shady windowsill. Meanwhile, if you are a fragrance lover, neos deliver some of the best in the orchid family. If you're looking to cultivate a new obsession, a single wind orchid today can cost anywhere from $15 to $100,000.

Orchid Details

~ Place of origin: Japan, China, Korea
~ Blooming season: late spring to early summer
~ Flowers last for: one week if warm; up to three weeks if kept cool and shaded
~ Plant size: tip to tip leaf span is usually 8 inches on standard varieties; there are also much smaller "bean leaf" varieties

~ Flower size: clusters of three to seven flowers; flowers are usually 1 inch across and 3 inches tall or more
~ Fragrance: changes depending on the time of day and variety, but can be described as jasmine, honeysuckle, coconut, orange blossom, vanilla, butter cookie, or butterscotch

— Plant habit: monopodial, bunching, easily trained and
displayed

Orchid Needs

— Light: shady; low to moderate light for green varieties,
higher for yellow-leaved varieties
— Temperature range: From mid-30s to mid-50s°F in winter
months, 50s to 90s°F in other seasons
— Humidity: Appreciates 75%–85%, but will cope with lower
— Water: high April through September, highest in June (and
throughout active growing and blooming season); much
lower October through May, with almost no water in cool
to cold winter months
— Growing medium: must be open, airy, and fast draining; many
growers prefer loosely packed long-fiber sphagnum moss,
but neos do well in medium-gauge bark and perlite, too
— Fertilizing schedule: half-strength balanced fertilizer with
micronutrients applied weekly during months of active
growth
— Seasonal changes: to flower well, neos must have a cool (in
the 50s°F), dry rest for several weeks (or a few months) in
winter; when necessary, misting or light watering every few
weeks is sufficient
— Special requirements: move a blooming Neofinetia into your
bedroom at night and let its fragrance inspire your dreams

Chapter 7

✣

The Science of Freedom and Charles Darwin's "Little Book on Orchids"

Charles Darwin (1809–1882) cut his botanical teeth traveling the world on the *HMS Beagle* for five years, spent almost a decade studying barnacles, upended Christian institutions worldwide with his theory of natural selection, and was subject to global ridicule and outrage—and then he turned to orchids. Darwin confided, "I am a gambler, & love a wild experiment," to Joseph Dalton Hooker, at the time the assistant director of the Royal Botanic Gardens, Kew, confessing that his work on orchids was as thrilling as any work he'd done.

The naturalist is of course best known for his first major work, *On the Origin of Species* (1859). However, he published several other volumes, many of which theorized on orchid growth and evolution in novel

Charles Darwin, c. 1854.

ways. Immediately following *Origin*, Darwin published *The Various Contrivances by Which Orchids are Fertilised by Insects* (1862), which he called his "little book on orchids." In it, he exhaustively detailed orchid pollination mechanisms, the lot of them performing ingenious sexual subterfuge upon insects. Orchids, he argued, are "universally acknowledged to rank amongst the most singular and most modified forms in the vegetable kingdom," and thus were the perfect follow-up case for his theory of natural selection.

Darwin's fascination with orchids was piqued by those he saw around him—the flowers he found most "multiform & truly wonderful & beautiful," a love that began more than two decades before he wrote *Various Contrivances*. Orchids were more than mere static plants; they exhibited power and movement and acted upon strong desires in concert with their pollinators. He felt since his time on the *Beagle* in the Galapagos in 1835, "At last gleams of light have come, & I am almost convinced (quite contrary to opinion I started with) that species are not (it is like confessing a murder) immutable." Even if the planet and everything on it were created by God, everything also evolved.

In *Various Contrivances*, fodder for his first chapters came from the native British orchid species he had studied for much of his life. He encountered early purple orchids (*Orchis mascula*), striking pyramidal orchids (*Anacamptis pyramidalis*), musk and frog orchids (*Herminium monorchis* and *Dactylorhiza viridis*), early spider and bee orchids (*Ophrys sphegodes* and *apifera*), and monkey orchids (*Orchis simia*). Echoing his notoriously licentious grandfather Erasmus, Charles described how "moths and butterflies perform their office of marriage-priests" for terrestrial European orchids, and how hard-working "humble bees" could be seen copulating with field orchids across England.

Darwin was keen to study tropical orchids too. He borrowed several almost-blooming plants from Hooker and returned them to Kew's collection once he finished his investigations. Tended in a small glass lean-to behind his home, Darwin watched many orchids' inflorescences develop and their blooms crack open—most orchid flowers achieving

Marbled white butterfly (*Melanargia galathea*) on pyramidal orchid (*Anacamptis pyramidalis*) at Darwin's Orchis Bank, now Downe Bank Nature Reserve in Kent, England.

full size and scent within three days of opening. He also had a reliable worldwide network of scientific and botanical colleagues more than willing to share their plants and findings with him. Some of the greatest defenders of *Origin of Species*, including American confidante Asa Gray at Harvard University, had worked with orchids for years and offered up additional examples of how the flowers perfectly fit Darwin's theories of natural selection.

Darwin considered adding his collected information on orchids to a later edition of *Origin of Species*, but the examples were too numerous— no, he decided, the work on orchids deserved its own book. It was a three-hundred-page tome to the beauty and the oddity of the orchid family (and perhaps no less as a professional love letter to Hooker and Gray), but most of all, it was proof that natural selection was at work in every tiny crevice of nature. Once published, Gray wrote, "if the Orchid-book had appeared before *Origin*, the author would have been canonized rather than anathematized." Gray may have been right. It took Darwin's work on orchids to ultimately win over the scientific community to his earlier hypotheses regarding evolution.

But why choose orchids? Darwin could have picked many other flora

or fauna, given that he had extensive experience with earthworms, finches, tortoises, pigeons, dogs, and barnacles. As a naturalist, he didn't even truly specialize in botany—he called himself a "Botanical ignoramus" to Hooker as late as 1844. We might guess that one of his reasons was the sheer proliferation of orchids on Earth—he knew that as

Darwin as an ape. *The Hornet* magazine (1871).

the second largest plant family, orchids spanned the globe. He also had firsthand knowledge from British orchids that the flowers had countless odd relationships with their pollinators. Lastly, Darwin proved canny in following trends—piggybacking on the rage for orchids in Victorian times was the perfect way to launch an assault on professional "monkey book" naysayers, and pull the public onto his side as well.

Darwin may have had other personal reasons for picking up with orchids where he left off in *Origin*. As biographer Janet Browne has argued, Darwin was most likely referring to a favorite family spot, Orchis Bank, in the dreamy conclusion to his previous book. He writes,

> It is interesting to contemplate an entangled bank, clothed with many plants of many kinds, with birds singing on the bushes, with various insects flitting about, and with worms crawling through the damp earth, and to reflect that these elaborately constructed forms, so different from each other, and dependent on each other in so complex a manner, have all been produced by laws acting

Fly orchid (*Ophrys insectifera*) at Darwin's Orchis Bank, now Downe Bank Nature Reserve in Kent, England.

around us . . . and that, whilst this planet has gone cycling on according to the fixed law of gravity, from so simple a beginning endless forms most beautiful and most wonderful have been, and are being, evolved.

Darwin's "entangled bank," his Orchis Bank, is now Downe Bank Nature Reserve in Kent, still today habitat for eleven species of orchids. Even Darwin's children reminisced about the place fondly. His son Francis remembered it as a quiet spot where grew "Cephalanthera and Neottia under the beech boughs," and his daughter Henrietta recalled Orchis Bank as a grassy terrace high above the valley within which "bee, fly, musk, and butterfly orchises grew."

Orchids, beyond their locality and sexual complexity, were also a good choice for Darwin because of their unique form. Many orchids have a "column," made of fused male and female parts. Yet, against assumption, most orchids do not regularly pollinate themselves in nature; the rostellum, another apparatus unique to orchids, is plant tissue that separates the anther and the stigma, preventing self-fertilization. Orchids' ornate structures actually promote adaptive outbreeding, guiding an insect up its labellum and stigma to ensure pollination by a flower the insect has previously visited. Darwin enthused, "An examination of [orchids'] many beautiful contrivances will exalt the whole vegetable kingdom in most persons' estimation."

One of the orchids that Darwin perhaps unwittingly exalted in his book was *Angraecum sesquipedale*. The gorgeous three-foot-tall plant from eastern Madagascar produces multiple flowers at a time. Its waxy blooms start out green but mature to a crystalline creamy white; the flowers are often eight inches across, and its rather shocking spur is an additional thirteen inches long. The orchid has been called the comet orchid, the king of angraecums, the foot-and-a-half-long orchid, and the Christmas star orchid. It is now best known, however, as "Darwin's orchid" because the scientist successfully surmised that the flow-

Angraecum sesquipedale. Warner, Williams, and Moore, *Orchid Album* (1897).

T.W.Wood del.   M & N Hanhart,imp.

SPHINX MOTH FERTILIZING ANGRŒGUM SESQUIPEDALE IN
THE FORESTS OF MADAGASCAR.

er's very long nectary could only be pollinated by an insect—likely a moth—with a very long proboscis. In January of 1862, Darwin wrote to Joseph Hooker, reporting that a friend sent him a box full of "the astounding *Angraecum sesquipedalia* [*sic*] with a nectary a foot long. Good Heavens what insect can suck it." He wrote again that same week, "What a proboscis the moth that sucks it, must have! It is a very pretty case." Indeed it was a pretty case of orchid-pollinator coevolution, but a moth the size that Darwin predicted—one with a tongue almost a foot long—wasn't found until 1902, twenty years after Darwin's death. And it wasn't until 1992 that the eight-inch-wide hawk moth of Madagascar, *Xanthopan morganii praedicta*, was confirmed to be feeding from *Angraecum sesquipedale*.

However popular *Angraecum sesquipedale* might be in preserving Darwin's legacy in orchid form, for Darwin himself, catasetums from Central and South America were "the most remarkable of all orchids." This truly odd genus has something that other orchids do not: the ability to, at high speed, eject its pollen packet onto its unsuspecting pollinator, affixing its gametes to a bee's head or back with nothing short of quick-drying super-sticky glue. First, a male catasetum flower—

(facing page) Thomas Wood, "Sphinx Moth Fertilizing Angraecum sesquipedale in the Forests of Madagascar," *Quarterly Journal of Science* (1867), published well before proof was found.

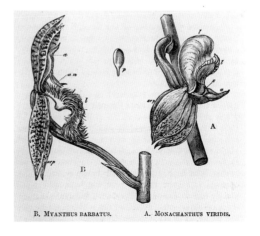

Male and female forms of catasetum species in Darwin's *The Various Contrivances by Which Orchids Are Fertilized by Insects* (1862).

B. MYANTHUS BARBATUS. A. MONACHANTHUS VIRIDIS.

because, not to be outdone, catasetum species produce male, female, and hermaphroditic flowers—emits a fragrance attractive to large euglossine bees. Entering the flower, the bee triggers the catasetum's propulsive device, and thus clobbered with pollinia, exits the flower. Remembering its abuse, it does not visit a male catasetum flower again. Female catasetum flowers, though—looking and smelling different— continue the orchid's game by luring the bee to enter and deposit the pollen on its stigma. As he was finishing the orchid book, Darwin wrote Hooker, "If you can really spare me another Catasetum, when nearly ready, I shall be most grateful. Had I not better send for it? . . . The case is truly marvelous . . . A cursed insect or something let my last flower off last night!" He later joked with Hooker that he needed to "rob your hot-house" for more orchids. To another friend he recalled describing catasetums' projectile pollinia to an incredulous colleague, who retorted back, "Do you really think I can believe all that?" Witnessing the propulsive action of *Catasetum tridentatum* (now *Cstm. macrocarpum*), Darwin found the orchid's way of attaching pollinia to the backs of bees "appears to me one of the most wonderful cases of adaptation which has ever been recorded."

John Lindley, Britain's foremost orchidologist at the time, wrote about Darwin's work on identifying the three separate types of catasetum flowers that "such cases shake to the foundation all our ideas of the stability of genera and species, and prepare the mind for more startling discoveries than could have been otherwise anticipated." Darwin acknowledged that the endless variations on a theme in orchid-pollinator relationships didn't seem to make logical sense. How is it that natural selection could occur in such a constant wilderness of forms? "In my examination of Orchids, hardly any fact has struck me so much as the endless diversities of structure . . . [but] this fact is to large extent intelligible on the principle of natural selection." The naturalist successfully argued that orchids' utter lack of foresight, their unbridled randomness, was the best evidence yet for his big theory.

Nature's lack of intelligent design proved just that: no omniscient creator was at work. "It is hardly an exaggeration to say that Nature tells us, in the most emphatic manner, that she abhors perpetual self-fertilization," and thus, change is built into life on planet Earth.

Perhaps more than their fascinating physical characteristics, Darwin found himself treading into philosophical territory with catasetum orchids: "How then does Nature act? She has endowed these plants with, what must be called for want of a better term, sensitiveness, and with the remarkable power of forcibly ejecting their pollinia even to a considerable distance." He goes on to say that all genera in the *catasetinae* subtribe, as well as many vandas, exhibit "sensibility."

> The study of these wonderful and often beautiful productions, with all their many adaptations, with parts capable of movement, and other parts endowed with something so like, though no doubt different from, sensibility, has been to me most interesting. The flowers of Orchids, in their strange and endless diversity of shape . . . appear to us as if they had been modelled in the wildest caprice . . .

Although he falls short of openly comparing the "sensibility" of orchids to humans—instead he finds plants similar to insects, fish, and other animals—he uses human terms to describe them. The words *sensibility, sensitiveness, sensation,* and *sensitive* occur more than forty times throughout *Various Contrivances* to describe orchid labella (lips), rostella, pollinia, anther cases, and the orchid flower that encompasses these reproductive organs.

This is important because *sensibility* and *sensitiveness* were watchwords in the Victorian age. More than being merely a stimulus to touch, sensibility had emotional and moral weight. Displaying appropriate sensibility was a way of proving one's gentility, one's good breeding and moral compass, a ladder to the upper echelons of society.

From a photograph by W. Ellis.

ANGRÆCUM SESQUIPEDALE AND NATIVE FERNS

Slaves carrying man past *Angraecum sesquipedale*. William Ellis, *Three Visits to Madagascar* (1858).

Interestingly, Darwin suggests that he detected the trait in orchids in a time when popular comportment in England and America demanded it, especially of white women. The fact that orchids not only moved but also showed sensitivity radically punctured the animal/plant divide—much as *Origin* had punched holes in the human/animal divide—and further threatened to upend racial hierarchies. This was in part sheer science, of course, but Darwin had also been raised in a staunchly abolitionist family. After personally witnessing the horrors of slavery in Brazil while on board the *Beagle*, his hatred of the institution gained strength. He wrote to Gray in 1861—and after asking whether the US botanist could send some information about cypripedium and spiranthes orchids—saying he wished to God "that the North would pro-claim a crusade against Slavery . . . how I should like to see that greatest curse on Earth Slavery abolished." After all, if orchids had sensibility, couldn't all races of people qualify as human? He had done his best to prove as much in his previous book.

The fallout from *Origin* continued to take its toll on the naturalist. Since its publication, Darwin had been ill with physical—and likely psychological—ailments. He suffered from insomnia, a stammer, nausea, abdominal pain, heart palpitations, and eruptions of eczema. The stress of providing proof of natural selection—against intelligent design—often left him prostrate and unwilling to engage in a social circle other than through the post. His pain wound through *Various Contrivances*, as he wrote to friend Sir Charles Lyell: "I am very poorly today & very stupid & hate everybody & everything . . . One lives only to make blunders—I am going to write a little Book on orchids & today I hate them worse than everything."

But Darwin had many allies and friends supporting him. Once the book was published, Asa Gray wrote to Darwin that he was "amused to see how your beautiful *flank*-movement with the Orchid-book has nearly overcome" all naturalists who balked at *Origin of Species* three years prior. Darwin was tickled at this and complimented Gray back, "of all the carpenters for knocking the right nail on the head, you are the very best: no one else has perceived that my chief interest in my orchid book, has been that it was a 'flank movement' on the enemy." The war metaphors were apt, given *Various Contrivances* was published in 1862, the dog days of the American Civil War. Darwin had carnage on his mind while writing *Origin*, too, in that it contained concepts that slipped from the human into the animal world—the words *battle, war, slave, attack, combat, survival,* and *struggle* were in frequent use in the book.

Darwin and Gray well knew that *Origin* had been a deeply subver-sive text for American society, and for colonial powers worldwide, far beyond its possible fracture of a monotheistic worldview. As author and professor Randall Fuller has argued, the crisis that stemmed from

Origin was, astonishingly, more about its social and political moment than about its actual content. When it was published in late 1859, slavery was still practiced in the United States, Brazil, the Congo, China, and Malaysia, and agricultural serfdom was the rule throughout Russia. Dedicated to the antislavery cause, Darwin corresponded with far-flung friends about these events and wondered how he might assist in slavery's demise.

Among the people Darwin first sent the book to was Gray at Harvard, reigning as America's best-known botanist. For years, Darwin—humble, politic, and knowing darn well that he was going to need advocates for his work across the Atlantic—plied Gray with questions about plant distribution problems in North America, and particularly about the eastern US-East Asia puzzle of the similarities of many species. These led Gray to think that much about plant geography seemed explainable only by some mechanism of descent through modification. So Darwin had a staunch advocate in Gray.

In 1859—also the year that gorilla specimens first arrived in America from West Africa—Gray passed his dog-eared copy of *Origin* to abolitionist and social reformer Charles Loring Brace. Gray knew what he was doing, giving his copy to Brace, a devoted anti-imperialist and antislavery advocate. Because *Origin* hints that all humans are related through a common primate ancestor, it very quickly became a powerful tool for abolitionists. Slavery relied upon a theory of radical racial difference—it was long argued that black people were not fully human, and therefore could be enslaved, owned, and treated like chattel. *Origin* destabilized that notion at its very core—through human biology.

Brace then passed that same copy to Franklin Sanborn and writer and naturalist Henry David Thoreau. Sanborn was a Massachusetts journalist, teacher, and avid abolitionist, as well as a critical member of the Secret Six, an organization of men who financially supported John Brown in his violent campaign to end slavery by any means necessary. In the 1850s, John Brown attempted to build an armed slave revolt across the South. In the summer of 1859, with a band of men, Brown had attacked the federal armory at Harpers Ferry in Virginia (now West Virginia) to collect guns for his cause. He ultimately surrendered and was tried, convicted, and hanged for murder, slave insurrection, and treason. His execution occurred just two weeks before *Origin* hit US shores. And so slavery, radical takeover, primate descent, death, and evolution were in hot debate internationally, intermixed in a flood of violence reported in the media.

The Secret Six had not known the full details of Brown's plan, though Sanborn had known the most, serving frequently as Brown's contact for the rest of the group. It seems important that Sanborn was perhaps the third person in America to read *Origin of Species*, and just after the national crisis at Harpers Ferry. He was a man looking for ways to further support the antislavery cause and was clearly part of an international network of professional naturalists who agreed with his mission.

With all of this in mind, imagine word of the continuous slaughter of the US Civil War reaching Darwin as he was making his final edits to *Various Contrivances*, with the finished book arriving on US shores in the summer of 1862. The fate of the Union forces was still very much in question, as the Battle of Shiloh that was waged just weeks earlier proved one of the bloodiest of the Civil War. Thus the stakes undergirding that little orchid book, replete with the scientific facts of orchid reproduction, modification, and natural selection, were far higher than they seem at face value. The critiques of *Origin of Species* that Darwin took most seriously were the ones lodged by other serious natural philosophers—people open to his ideas, but not yet swayed, for lack of primary evidence. With the orchid book, Darwin was back on the scene, and he brought overwhelming evidence for his theories. Writing from Orchis Bank, Darwin provided more staunch fodder for antislavery advocates around the world.

Origin of Species and *Various Contrivances* are ingenious scientific works, but ones that are made all the more amazing by their dates of publication and the community of abolitionists who first received them. The books threatened to not only decenter religious worldviews—the way they are pigeonholed today—but also overthrow multiple patriarchal and racist power structures at once. They depicted a constant struggle of life and death, endless competition, and ingenious biology. While *Origin* made the first and hardest blow, Darwin's little book on orchids further destabilized racist ideologies, providing proof of a grand scientific revolution that reinforced a larger age of political revolutions.

ANGRAECUM SESQUIPEDALE

*A*ngraecum sesquipedale is popularly grown in greenhouses, but grows to several feet high, and is therefore too large for most indoor orchid collections. *Angraecum didieri* and *rutenbergianum*— miniature cousins of *sesquipedale*—are good under lights and on windowsills; they are also endemic to Madagascar and have a similar shape: a crisp white star with long, elegant nectary.

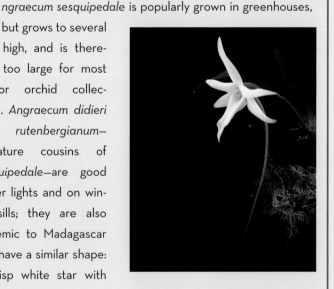

Coulanges/Shutterstock.com.

Orchid Details

— Place of origin: east coast of Madagascar
— Blooming season: early winter through early summer in the
   northern hemisphere
— Flowers last for: weeks to months
— Plant size: 3 feet tall or taller
— Flower size: 8 inches across, 21 inches long including its
   spur
— Fragrance: spicy nocturnal fragrance, similar to lilies or
   nicotiana
— Plant habit: monopodial—grows tall but erect and thin

Orchid Needs

— Light: as an epiphyte, it grows in trees with some shade, so
   dappled or bright indirect light is best
— Temperature range: warm to hot grower; no lower than
   60°F nights, with highs in the mid-80s°F
— Humidity: 80% ideal, will cope with less
— Water: heavy most of the year, heaviest in summer (8
   inches per month), about half that much water in winter
   months
— Growing medium: grow mounted or in well-drained baskets
   with coarse fir bark, large perlite, and charcoal
— Fertilizing schedule: use balanced orchid fertilizer at one-
   half strength when actively growing
— Seasonal changes: water and fertilize less in winter
— Special requirements: strong air movement, avoid
   repotting—only repot when you see new root tips

CATASETUM MACROCARPUM

Also known as the monk's head, monkey goblet, or large-fruited orchid, you may find male or female flowers on your *Catasetum macrocarpum*. The light and humidity the plant is subject to generally determines its sex for the blooming season: high light and low moisture tend to induce female flowers, whereas low light and high humidity produce male blossoms. While the male form is often preferred for its color and propulsive pollinia, you might prefer the female (pictured here)—they resemble little green aliens.

Rui Santos/Shutterstock.com.

Orchid Details

~ Place of origin: Venezuela, Brazil, Trinidad, Tobago, Suriname, Guyana
~ Blooming season: throughout the year, highest July through January in northern hemisphere

- Flowers last for: one to two weeks if female; once the pollinia is triggered in male flowers, they fade quite quickly
- Plant size: large, arching leaves and thick pseudobulbs create an orchid that is several feet wide and tall
- Flower size: 3 inches wide
- Fragrance: caraway, mint, citrus, pepper
- Plant habit: sympodial; found growing epiphytically on trees, leaves arch gracefully away from the trunk

Orchid Needs

- Light: moderately bright; no direct midday sun
- Temperature range: hot grower; no less than 55°F, highs in the high 80s°F
- Humidity: 75% or more all year
- Water: moderate water April through November, almost no water January through March
- Growing medium: mounted or in well-drained baskets with fir bark, large perlite, and charcoal
- Fertilizing schedule: heavy feeder; use full-strength balanced orchid fertilizer throughout the growing season; do not fertilize in autumn or winter
- Seasonal changes: in the northern hemisphere, leaves drop in autumn; when this happens, water only to prevent pseudobulbs shriveling, and resume watering in April when new growth is 4–5 inches long
- Special requirements: strong air movement, dry winter; many growers recommend annual repotting in spring as new growth begins

PART III

The Business
of Beauty

Chapter 8

⌒ℐ

Itinerant Orchids, Enslaved People

*V*anilla planifolia is the only flower out of the tens of thousands in the larger orchid family to produce an edible seed product that has sustained its global trade for almost two hundred years. In the wild, vanilla orchid seeds develop inside of fleshy pods, waiting for the precise moment to burst into a forbidding and dangerous jungle. These seeds—thousands in each successfully pollinated vanilla orchid bean—have no shell, no defenses, and no nutrients to see them through the sprouting phase. Instead, they depend upon the natural environment to provide a perfect mix of soil fungi, sun and shade, water and heat to ensure their growth.

A similarly perfect mix of environmental factors for growth was not the good fortune of Edmond Albius (1829–1880), a twelve-year-old

Vanilla planifolia.

enslaved boy in 1841 when he became the first person to successfully hand-pollinate the vanilla orchid. His horticultural skill on the island of Bourbon (off the coast of Madagascar, today known as Réunion) allowed for a global boom in vanilla, revolutionizing the industry.

But if we back up—back way up to seventy-eight million years ago—we find vanilla in the late Cretaceous period amongst the dinosaurs. Vanilla made it through the extinction event that killed 75% of life on Earth, and eventually diversified into more than 120 species. Vanilla vines are now endemic throughout the tropical Americas, Africa, Madagascar, India, and south Asia. Yet only three species are used for vanilla flavoring. *Vanilla planifolia*, native to Mexico, is used for 95% of the world's extracts. *Vanilla pompona* and *Vanilla x tahitensis* (a primary hybrid of *planifolia* and *odorata*) round out the rest, and contribute to the vanilla trade in other ways, chiefly for perfume. While the vanilla flower is simply sweet, its ripe seed pod opens a universe of aromas. Food scientists describe it as floral, balsamic, earthy, woodsy, nutty, dusty, smoky, rummy, resinous and spicy, and with hints of the barnyard, hay, tobacco, almonds, raisins and prunes, bananas, and cotton candy. Pure vanilla holds more than three hundred volatile compounds.

Vanilla is an orchid that has benefited from, and been decimated by, global trade. Originating in central Mexico, the indigenous Totonac peo-

Diego Rivera, mural of Totonac culture at the Palacio National in Mexico City. Note vanilla vines and flowers at the center.

ple first cultivated the fragrant spice, likely beginning 2,500 years ago. Their foundational myth records that vanilla flowers and pods sprang from the blood of a princess who was sacrificed by her father and temple priests when her love for a handsome commoner ran afoul of tolerable behavior. The perfume of mature vanilla is said to be Tzacopontziza's sweet soul, and through her the vanilla orchid is forever sacred. Totonac culture is still rooted in the vine—if you visit Papantla, Veracruz, at the height of vanilla production in Mexico in early summer, you can attend a vanilla festival which has been held annually for hundreds of years.

Around 1248 CE, the Mexica people—later called the Aztecs—migrated into vanilla-growing regions of Mexico, and vanilla soon became an indispensable ingredient in their famed *xocolatl* drink made of cocoa beans, vanilla, chili, and corn. Emperor Montezuma II reportedly drank fifty cups a day to sustain his sexual stamina for his many wives. By the 1520s, Spaniard Hernán Cortés and his conquista-

Vanilla in Fray Bernardio de Sahagun's *Florentine Codex, Book XI: Natural Things* (1577).

dores colonized Mexico, and he is credited with introducing vanilla, chocolate, and many other foodstuffs to Europe.

For the next three hundred years, vanilla was a luxury item reserved for the very wealthy. Demand was high, and following much trial and error, several European colonial enterprises eventually exported vanilla vines to plantations throughout the Indian Ocean. Yet they still hadn't mastered how vanilla was pollinated, making large-scale production of it impossible.

On one of those plantations, on the island of Bourbon off the coast of Madagascar, a boy named Edmond was born to an enslaved maid, Melise, in the Sainte-Suzanne district. (Edmond's surname was added only after his emancipation in 1848.) Melise died in childbirth, and Edmond never met his father. As a young child, he was given to Ferreol Bellier-Beaumont to work at his nearby estate at Belle-vue. Bellier-Beaumont was known regionally as a discerning horticulturist who liked to experiment with edible crops and ornamental plants. Edmond proved a constant companion to Bellier-Beaumont in his gardens, and learned horticulture under the tutelage of his master. The boy was a careful observer of other crops, especially watermelons, being pollinated by insects and by hand in the plantation's gardens. In 1841, he invented a method to quickly and reliably pollinate vanilla flowers. His technique was to insert a small stick sideways into a blooming flower just under the rostellum, which separates the anther and stigma. By lifting the rostellum forward and upward, the anther was also pressed forward, depositing its pollen on the stigma. Bellier-Beaumont described later, "This clever boy had realized that the vanilla flower had male and female elements, and worked out for himself how to join them together." The orchid had evolved over millions of years to prevent such self-fertilization, but Edmond had devised a way around its botany.

Bellier-Beaumont quickly invited other landowners to his plantation for a lesson in vanilla pollination. Soon, the enslaved boy was escorted by carriage to all major landholdings in the region to demonstrate *le geste d'Edmond*, or the art of "orchid marriage" to other enslaved people. His practice also promptly traveled to other vanilla farming areas across the Indian Ocean and throughout Indonesia and Tahiti.

As with the invention of Eli Whitney's cotton gin in the Americas, while hand-pollinating vanilla allowed for incredible profits for owners, it also accelerated the growth of slavery and the destruction of flora throughout the Indian Ocean region. In particular, for more than a century, Bourbon had been a large slave plantation system for coffee and sugar, so the stage was set for the labor

ALBUM DE LA RÉUNION.

A. Roussin. del. et lith. d'ap. nature   1863.   Imp. A. Roussin. Ile de la Réunion.

EDMOND ALBIUS
Inventeur de la fécondation artificielle du vanillier.

Edmond Albius with vanilla vine. Antoine Roussin, lithographer, *Edmond Albius Inventeur de la Fécondation Artificielle du Vanillier* (1863).

force—technically free in 1848—to transition to vanilla. That year, the recently renamed Réunion exported 110 pounds of dried vanilla beans; by 1858, the number topped 6,600 pounds, and by 1867 the figure was twenty tons. At its apogee in 1895, Réunion shipped two hundred tons of cured vanilla to Europe. The island and the rest of the Muscarene island chain had already eclipsed Mexico as the world's greatest exporters of vanilla almost a decade before.

When freed, Albius chose not to pursue vanilla farming. His surname ironically, or perhaps symbolically, means "white"—we do not know if Albius chose the name, it was in reference to his free status, or perhaps

Vanilla aromatica in Etienne Denisse, *Flore d'Amerique* (1843–1846).

was a poor joke played by the white men in charge of the registry. Despite his horticultural knowledge, various forms of racism put very real limits on his ability to prosper on the small island. Unable to find an immediate place in this, for him, impossible society, he ended up convicted of robbery, serving time in prison, and dying in poverty. A newspaper in Réunion recorded his death in August 1880: "the very man who at great profit to this colony, discovered how to pollinate vanilla flowers, has died in the public hospital at Sainte-Suzanne. It was a destitute and miserable end."

Today, racism continues to haunt Albius's achievement. As journalist Tim Ecott has written, some people on Réunion continue to deny that Albius invented the hand-pollination of vanilla. In fact, historical documents prove the veracity of his creating the technique—including an archive of letters written by Bellier-Beaumont in 1855 to the governor of Réunion, Louis-Henri Hubert Delisle, describing exactly when and how Albius first performed *le geste* in an effort to convince the official to reduce or repeal Albius' prison sentence. A respected horticulturist himself, Bellier-Beaumont details how his "favorite boy" had been taught other plant fertilization methods, but that Albius alone had revolutionized the industry for Bourbon. Albius's former master appealed to the governor's compassion and sense of justice in the case of the young man condemned to hard labor. He wrote:

> For more than two years [Albius] has suffered in silence . . . [he] is just one of many slaves in our country, who was thrust into the wide world without proper preparation . . . [people] took advantage of his youth and inexperience . . . If anyone has a right to clemency and to recognition for his achievements, then it is Edmond. It is entirely due to him that this country owes a new branch of industry—for it is he who first discovered how manually to fertilize the vanilla plant.

His appeal was successful, and Albius was released that year. As the white plantation owner, Bellier-Beaumont could have stayed silent about Albius' conviction or taken credit for devising the pollination

Malagasy 25,000 franc banknote depicting vanilla harvest. *Prachaya Roekdeethaweesab/Shutterstock.com.*

technique since Albius wouldn't have had the ability to challenge him. That Bellier-Beaumont didn't adds credibility to his account.

Yet once vanilla made the short jump from Réunion to the northeast tropical rain forests of the French colony of Madagascar, the fourth largest island in the world made a mad dash to surpass all other regions' vanilla production. It too was primed for vanilla—holding a well-visited place on the spice route for centuries, the mini-continent was a behemoth among the pebble-sized islands of the Comoros and Bourbon islands. All of them had developed a lucrative trade in intoxicating fragrances and spices for more than a century—balms like frangipani, patchouli, ylang-ylang, lemongrass, cloves, ginger, basil, and pepper. But it was vanilla, introduced to Madagascar's shores in 1840, that easily became its most famous and bankable export. Slavery was not outlawed by its colonial administration until 1896, so its profits were enormous for the French Empire.

For better or worse, by the year 1900, the fragrant orchid was described as a "veritable magic wallet . . . of usefulness and delight" by Joseph Burnett, a lauded innovator of vanilla extract located in Boston. And yet before that, in 1874, German scientists had begun

VANILLA (Orchid and Bean)

Vanilla orchid and bean.
Artemis Ward, *Grocer's
Encyclopedia* (1911).

synthesizing vanilla flavor, vanillin, by using coniferin—a component of pine bark. Some years later, a French chemist in 1891 extracted
vanillin from eugenol, a natural compound found in clove oil. Imitation vanilla competed with vanilla, both becoming ubiquitous as a
flavoring, as well as a flavor toner, smoothing and balancing tastes
in drinks and foods. Because of vanilla's expense and long production time, vanillin made from "flavor alternatives" makes up 97% of
the vanilla flavoring found in foods, sodas, perfumes, medicines, pet
products, rubber and plastic toys, and everything else we buy that
contains the scent.

The production of vanilla continues to be a long, slow process.
Each seed pod ripens separately, making harvesting a daily chore.
It is also very slow to cure. Scented vanillin does not accumulate in
the fruits until the fourth month after pollination and continues to

Vanilla mexicana as *Epidendrum vanilla* in Zorn, *Icones plantarum medicinalium* (1779–1790).

develop for the next three months. Yet, the pod isn't fully ripe until the ninth month. If you're thinking that the development of vanilla pods sounds a lot like the course of human pregnancy, you're not alone—overlapping names and references for the plant and human sex have been around for centuries. *Vainilla* is Latin for vagina, which itself also means "small sheath" or "pod." The link between vanilla and human reproductive organs was clear for the Totonac people, too—the orchid is divinely tied to humankind in their cosmology.

After finally harvesting the pods, growers "kill the beans" in numerous ways: they lay them out in the hot sun, dry them in an oven, scald them in boiling water, freeze them, expose them to ethylene, or scratch and sweat the pods. Those processing vanilla need to then dehydrate the beans. This requires further watchful care, because done too fast, it may scorch the crop, while drying them too slowly risks rot. Curing alone takes no less than two months—it develops vanilla's complex flavor and its shelf stability—but smaller and larger beans require different lengths of time, and so supervision of the crop never ceases. As an agricultural product, vanilla is a long-term investment; pollination to processed bean takes no less than a year and a half. For all these reasons, vanilla is thought to be the most labor-intensive agricultural crop on Earth.

Currently, vanilla crops ripen around the world throughout the year—in Mexico between March and May, in Madagascar from October to January, and in Tahiti July through October. Vanilla is a volatile crop; the price of one kilogram of vanilla (2.2 pounds) can swing from $25 to in excess of $500. Most recently, world trade in vanilla beans registers as a $1.2 billion dollar enterprise. Madagascar retains first place as the largest vanilla exporter in the world, with 56% of the market, followed by Indonesia. China, Mexico, and Papua New Guinea take the third through fifth places, and Réunion still retains a spot in the top ten, but barely. On the demand side, the United States makes up more than 40% of the market, while France purchases less than half that, and Germany, 10%. All other importing countries claim a global percentage of vanilla that is in the low single digits.

Vanilla may be the only orchid to sustain a global industry, but it is not the only edible orchid. Salep, made from the flour of dried tubers from dozens of genera of terrestrial orchids, is popular in ice cream and hot and cold drinks in Turkey, Greece, the eastern Mediterranean, and parts of the Middle East. While traditional harvesting techniques had been sustainable for centuries, with salep's growth in popularity has come increased commercial pressure on native orchid extraction throughout the region. Roughly thirteen orchids go into making one cup of salep, and the use of their tubers means whole plants—which only mature at seven or more years—are destroyed in the process. Very little salep cultivation is underway, so with the collection of wild orchids for salep comes the destruction of ecosystems.

Unlike salep, vanilla is a sustainable crop. It's cultivated best in a natural environment: a rain forest, climbing in native trees in part shade. As opposed to so many other agricultural products, clear-cutting is not now widespread in vanilla cultivation. The vine is also largely grown without pesticides, fungicides, or fertilizers, although many growers perfect their own blends of local organic mulches to provide low levels of continuous feed. Vines can live for decades when properly spaced in well-drained soil and the vines are not over-pollinated.

While the vanilla industry is growing around the world, native vanilla vines are on the verge of extinction in many places. In Mexico, it is estimated that there are only a few dozen indigenous vines left. Genetic diversity has been lost due to cloning a small number of cultivars, making vanilla susceptible to disease and drought. Rain forest destruction, excessive harvesting in the wild, and climate change are additional destructive factors. Vanilla's natural pollinator, a small, stingless euglossine bee, is also under serious threat.

Fusarium (or, as some call it, *the orchid pathogen that shall not be named*) can also devastate vanilla. Just like in private orchid collections, the fungus rips through orchids where good management is not practiced. There is no remedy nor cure for infected vines, so they must be removed and destroyed. Biosecurity and thoughtful culture greatly decrease the risk of infection.

The demand for vanilla still grows, but in many places it is an increasingly dangerous enterprise. Madagascar has suffered under climate change, hit by several cyclones and severe drought in the past decade, leading to vanilla booms and busts. Criminal enterprise and low pay also haunt the Malagasy vanilla farmer. Thieves steal beans before they are ripe and during the curing process; in turn, farmers sometimes harvest their beans early, bringing a lower price at market because the beans are smaller and less flavorful. One safety measure used since colonial times is to tattoo one's vanilla beans. Two weeks before harvest, farmers use a wooden tool with small nails to imprint their mark on the ripening beans. While tattooing doesn't prevent all theft, it does create a barrier for thieves who attempt to sell the beans at market. Similar difficulties in farming vanilla are found in Indonesia, Mexico, and China.

Vanilla farmers in these situations often take on the greatest risk within the wider industry, and yet make little in return—for instance, only 8% of sales go back to the farmer in Madagascar. So what can people in vanilla-consuming countries to do for the vanilla producers? One answer that has worked well for several other important foodstuffs is fair trade—importers and consumers ensuring that fair prices are paid to vanilla producers. Coffee and chocolate have developed robust fair-trade systems in recent years; tea, bananas, wine, cut flowers, cloves, and cinnamon are gaining ground as well. Although vanilla was added to the Fair Trade Certified portfolio in 2005, the availability of fair-trade vanilla still remains low. Fair trade won't solve all of vanilla production's problems, but it could help networks of vanilla farmers build more stable lives and businesses. Another answer for consumers is to simply buy more real vanilla rather than its imitation. The next time you bake cookies, think of the history and the promise of the vanilla orchid. Its bean produces one of the most complex flavors in the world, expressing hundreds of organic components in every bite. Vanilla is an ancient survivor, but that doesn't guarantee it will remain with us forever.

VANILLA PLANIFOLIA

If you're inspired to grow a vanilla vine of your own, it does make an attractive houseplant. It does not usually bloom in a home environment, but orchid lovers grow *Vanilla planifolia* for its glossy leaves, twining stems, and trailing habit. Just don't expect them to flower, let alone provide you with

aromatic pods for cooking. Because of these limitations, many growers opt to cultivate the beautiful, variegated lime green-and-white variety *Vanilla planifolia* 'Albomarginata' (picutred here). Its leaves are more delicate than those of most vanilla vines, and its cascading vines add a touch of translucent filigree to any sunny window. And if you live in an area with no chance of frost, you may want to plant a vine under a tree that is well-watered throughout the year.

Orchid Details

— Place of origin: Mexico, West Indies, Costa Rica, Guatemala, Central America
— Blooming season: can bloom at almost any time of year, but each vine blooms only once per year, and only after it is ten years old; do not expect blooms in a home environment

— Flowers last for: one day, but a set of flowers on a vine can bloom in succession for several weeks

— Plant size: a small stem cutting will grow to several feet in a matter of years; flowers require the plant be 12 feet tall or longer; grows to 100 feet tall in the wild

— Flower size: 2 inches

— Fragrance: sweet (the vanilla scent only develops with the seed pod)

— Plant habit: climbing monopodial vine

Orchid Needs

— Light: light shade to bright shade, never full sun

— Temperature range: consistently warm at 65°F–90°F is best, but vines will usually adapt to lower temperatures

— Humidity: 80% or higher is best, but will cope with lower

— Water: water consistently and plentifully; do not let the soil dry out—but also don't let it remain waterlogged; thoroughly soak once or twice a week

— Growing medium: Vanilla is a hemiepiphyte, meaning it is terrestrial and epiphytic. Although vanilla wends through trees and gains water and nutrients from rain, its roots are in the soil. Potting mix can be denser than for most orchids, and include peat moss, bark, potting soil, leaf litter, and small perlite.

— Fertilizing schedule: vanilla is a heavy feeder—use half-strength fertilizer all year

— Seasonal changes: very little; thrives in consistently warm and wet environments

— Special requirements: most growers with the space create a sturdy support for the orchid to coil up and around

Chapter 9

Jane Loudon and Her Floriferous Press

At first blush, the life of novelist and garden writer Jane Loudon, born Jane Webb, seems fit for twenty-first century sensibilities. She was a science fiction author; she excelled at kitchen gardening and DIY; she agitated not only for women's rights but also women's pleasure and independence. And yet Webb Loudon died in the mid-nineteenth century, well before suffrage was an open struggle, before corsets and crinolines were shucked, before single women could lay claim to their own property, and what's more, their own minds and bodies.

Although the English author's life was brief (1807–1858), she pressed the boundaries of acceptable women's reading and hobbies far ahead

Flower prints and headdresses—as well as corsets and crinolines—were in style in the mid-1800s. *Godey's Fashions* (1856).

of her time. Pages of botanical instruction, gothic horror, and fantasy flowed from her pen. In her youth, she had a formal education, and toured Europe, learning several languages. But Webb was soon without family—she was only twelve when her mother died, and her father passed on when she was seventeen—and she was therefore utterly self-reliant in early adulthood. Yet these years set the tone for her work to come. She urged women to seek their own interests, telling her readers that "our chances of being happy decrease in proportion as we depend upon others for our enjoyments." In her horticultural books, she included scientific detail and stunning prints of orchids, providing ready access to art for women who were literate but frugal and did not have the luxury of a formal education. Laying aside strict Victorian propriety, she advised readers in how to efficiently perform their domestic chores and in the art of living a meaningful life. In a time of manners and restraint for women, she enthused at the sight of a woman on horseback, "I like to see a lady ride well, and fearlessly."

Jane Loudon c. 1820s.

Loudon also popularized indoor and outdoor gardening—including cultivating orchids—for women through a series of groundbreaking handbooks. She offered how-to guidance on growing popular and rare orchid genera, including dendrobium, epidendrum, cattleya, stanhopea, and catasetum. In a time of very poor orchid advice—when it was widely thought that all orchids should be set in swampy, hot greenhouses—Loudon wrote from careful observation and experience that many epiphytic orchids should be grown in a cracked coconut, "half-filled with moss, from which the roots hang down," tied to the rafters of a greenhouse and set in a shady spot with plenty of fresh air. She quietly promoted cultivation of some of the most sexually provocative orchids while underlining larger efforts at women's struggles for equality and education.

Her first book, *Prose and Verse*, published in the hope of securing an income as a writer, declared her romantic delight in nature. The book reveals a young scholar and multilinguist: it contained translations of Cervantes and reworkings of classics from French literature and Greek myth. Her teenage pen was full of gloomy ruins, romance, and maidens in beautiful gardens "enameled with flowers." In some poems, Loudon

*Dendrobium nobile. Paxton's Magazine
of Botany* (1840).

herself is a sad flower. "Lines Addressed to a Rose" mourned, "Poor lingering relic of far happier hours, Why dost thou droop to earth thy languid head?" Yet, along with mourning and winter's "thickly pelting storm" come a new season, and Loudon, the industrious, optimistic rose "spring[s] a phoenix from its parents' tomb."

Loudon's next work, *The Mummy!*—she was now in genuine necessity of paying her own bills—was a science fiction epic released anonymously at the direction of her publisher when she was just twenty years old. The work is comparable to Mary Shelley's *Frankenstein* in that it is a foundational text in science fiction. But where Shelley's vision of the world was atheistic, pessimistic, and cried out for social change, Loudon underlined the presence of a universal divine power and dreamed of a futuristic England that after many political revolutions had returned to an absolutist monarchy: an exclusively matriarchal one led by young unmarried queens. Her image of twenty-second-century London is exquisitely steampunk, complete with automaton birds, robot physicians, stunningly prophetic mechanical milking and espresso machines, and mail delivery by cannon balls. The book also plays with deeper concepts of nature, and finds solace, beauty, and personal identity in growing plants. Loudon dreamed about the politics as well as the science and verdant promise of the future.

Among early readers of *The Mummy!* was John Claudius Loudon (1783–1843), a famous garden designer. Thinking the book's author male, he contacted the publisher to ask to be introduced to the writer. He met Jane in February of 1830, and by all accounts, was immediately smitten with the woman twenty-four years his junior; they were married seven months later. Jane and John soon kept a lively literary circle of friends—including Charles Dickens's family—and Jane especially enjoyed the company of artists. By 1832, they had a daughter, Agnes. Like her parents, she too would write, publishing several stories.

Reviewers of Jane Loudon's first gardening books assumed that all of her training and floral inspiration were gleaned from her husband.

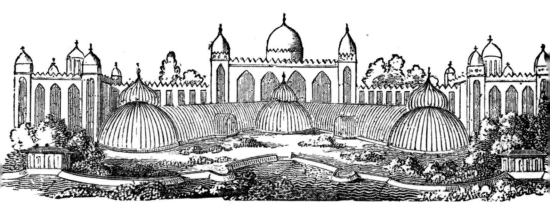

Inspiring to many future architects, this conservatory "of magnificent form" was included among several designs in John Claudius Loudon's *An Encyclopedia of Gardening* (1828).

While Jane picked up horticultural knowledge assisting John in his publications, the bulk of her botanical education came from the Father of Modern Orchidology himself, John Lindley. Newly appointed as a professor of botany at the University of London, Lindley gave a series of lectures on botany, and specifically orchids, in the early 1830s that Jane attended. However, while she learned botany from men, she had clearly appreciated nature, and in particular, exotic flowers, from young age. For Jane's virtuous Queen Elvira in *The Mummy!* "Beautiful flowering exotics decorated the [castle's] pavilion . . . and the balmy air that fanned their blossoms, seemed loaded with sweets . . . [it] had the appearance of a fairy palace. . . . There, vases filled with the rarest exotic flowers, shed sweet fragrance through the air."

Our story will focus on Jane, but John in no small way revolutionized orchid growing, too. He is best known for inventing his own horticultural style, one he called "gardenesque." Prescribing taxonomically grouped flowers and shrubs within formal grids in the garden, nature was subject to scientific education, aesthetics, and experimentation in his designs. But it is for his early and continuous work on greenhouses (also known as glasshouses, hothouses, or stoves) that orchid growers must thank him since it opened the door to the flowers' cultivation in cold climates. John began writing about hothouses as early as 1805, and after two more texts on the topic, published his wildly influential *Greenhouse Companion* in 1824. His principles of greenhouse design, construction, and maintenance were used by the nineteenth century's best-known architects, including Joseph Paxton in his famous Great Conservatory and Lily House at Chatsworth, and in the single most impressive glasshouse of the century, the magnificent Crystal Palace for the London International Exposition in 1851. Other architects' designs for greenhouses at the Royal Botanic Gardens at Kew also stemmed directly from John's teachings.

The Loudons kept their own gardens, but as writers, not landed gentry, theirs were nothing so elaborate as those they often wrote about. They lived in Bayswater, West London, in a home John had earlier built as a "semi-detached" townhouse to house himself in one residence and his mother and sisters in the other. In addition to setting a new standard for middle-class urban living, the house boasted a domed glass conservatory facing the street. The house and its grounds, also serving as the Loudons' editorial office and test gardens, were alive with activity day and night. The couple grew more than two thousand species of plants on site as bedding plants and in experimental beds.

Jane both wrote about her husband's greenhouse plans for a wider audience and gave additional advice in how to use the sunny spaces to a gardener's advantage. She assured those who might be trepidatious that for tropical greenhouse plants, "we must imitate as well as we can their native climate: that is, the degree of heat, light, and moisture they have been accustomed to in their native country, together with the air and the soil." Paying close attention to one's orchids would round out one's dreams of an exotic oasis. "All that is wanted to give an interest in any subject is, a sufficient degree of knowledge respecting it to be aware of its changes, and our own natural love of variety will do the rest."

Speaking to middle-class novices and women well above her station,

Jane advised that while experienced and serious gardeners keep no less than sixteen different plant houses for different climates, the careful orchid keeper could get away with four: a "dry stove" kept at 65°F–84°F, a "bark stove" kept at 60°F–80°F, a "forcing house" that might have a wider range of temperatures, and lastly, an exclusive "damp stove, or orchidaceous house." For this, she suggested temperatures of 70°F–90°F during the day, with high humidity. An additional conservatory attached to one's house, "lofty and architectural," was also a great addition to any country home. In reality, due to their finances, while the Loudons did enjoy a conservatory, they only had one hothouse, and it was devoted to orchids and tropical plants.

John's success with greenhouse design led quickly to his coveted seat as one of Britain's best-known gardening experts. Founding *Gardener's Magazine* in 1826 and running the show for almost twenty years, he also published encyclopedias on trees and horticulture. But British gardening magazine editorials were often vicious. Rival magazines would accuse each other of the theft of horticultural images and information, each magazine clamoring for the first printed evidence of new orchid species. Attendant with the premiere of new orchids in the best magazines were colored plates, illustrating the shocking—even lurid— flowers and foliage. These plates required enormous personal investment from editors up front, thrusting John and others to the verge of bankruptcy for decades.

The surprisingly cutthroat world of Victorian gardening magazines was all the more difficult for Jane. George Glenny, a cantankerous editor of several competing garden rags, was just one of the louder voices expressing his rage at Jane's mere presence in garden publishing. He wrote in his 1834 *Horticultural Journal* "[Loudon's] old woman is a mischievous beldam . . . we hate old women at the best of times, but a lying old woman is abominable, and the sooner Loudon shakes the hag off the better." The "hag" in question was twenty-seven years old and the mother of a two-year-old at the time.

On the face of it, Jane's work toward gender equality seems modest at best. But over the course of her career, there is a palpable trajectory. One piece of evidence is that many of her books were published anonymously or pseudonymously before and after her marriage but eventually she became widely respected as a woman publishing for women. One of the most ridiculous examples of her forced anonymity was her 1838 *The Young Lady's Book of Botany: A Popular Introduction to That Delightful Science*. Think of it: Jane had to publish anonymously, yet use the persona of a male narrator in her text, to launch the idea that botany could be an acceptable pastime for women. As one might imagine, repeated assurances that the pursuit of botany for ladies "is one of innocence and unalloyed pleasure, combined with healthful exercise" abound in the book.

In this early work, Loudon took steps to correct rampant scientific errors in orchid understanding and care, pointing out that many types of orchids were epiphytes living upon trees, not parasites sucking the

Tolumnia guttata as *Epidendrum guttatum* in Loudon's *Young Lady's Book of Botany* (1838). We can tell Loudon's budget was scanty due to the quality of her early works' images.

life out of their hosts, as commonly understood. At times she presages Darwin in writing "the formation of the flowers of all the *orchidaceae* is so singular, and the mode of action of the organs upon each other is often so obscure, that much attention is required in the study of them so as to ascertain the tribes and genera to which they belong." And in one of her first gardening books—*Instructions in Gardening for Ladies* (1840)—published in a size that was perfect for carrying in one's gardening apron pocket, she offered advice that remains patent for orchid growers today: never overpot indoor plants of any kind; an airy greenhouse is a healthy greenhouse; never repot when in flower; and keep the greenhouse clean and free of dead leaves—overcrowding your plants invites pests and disease.

Loudon was so impressed by orchids in her early career that she lobbied her publisher to include a full-color plate of an equitant oncidium, *Epidendrum guttatum* (now *Tolumnia guttata*) in *The Young Lady's Book of Botany*. She writes admiringly,

> The orchidaceae are no less remarkable for their forms than they are admired for the elegance and rich beauty of their blossoms. Out of the most grotesque habit, flowers of matchless brilliancy of colouring are produced. The cattleya is as vividly coloured as any other flower whatever; while some others are as hideously marbled with dull and lurid hues as can well be conceived. Almost all the South American species are remarkable, either for their modes of growth, distorted figure, or beauty of flowers.

With tolumnia, a miniature orchid found in happy yellows, pinks, and pearly whites, Jane seems most impressed by it and other orchids' "distinct and remarkable . . . lip, or labellum, which is often lobed, and assumes a great variety of forms." She would be pleased that this tolumnia, first systematized in 1753, is once again popular with orchid growers.

By 1842, Loudon had overcome both her enforced anonymity and made herself a popular authority in botanical science. Significantly, she does not include details in her books that had become other women authors' stock and trade, namely flower mythology and medical use in the home, nor did she repeat the "language" of flowers and "hidden meanings" of bouquets one might receive from a suitor. Instead, she advises readers of *Botany for Ladies*, "Nothing is more natural than to ask the name of every pretty flower we see; but unless the inquirer knows something of botany, the name, if it be a scientific one, will seem only a collection of barbarous sounds . . . half the interest of new greenhouse plants is thus destroyed." She was sure that science would increase women's love of flowers, not diminish it.

While expanding scientific learning for ladies, she also took the dicey step of openly describing flower pollination. She intoned that flower sex was key to understanding botany in any real way. She teaches, "The beautifully colored parts of flowers are the least important; and that, as they only serve as a covering to the stamens and pistil, which are designed for the production of seed." The anonymous Loudon performs a high-wire act in her description of flowers' sexual organs, for instance, "the ovary is juicy and succulent when young. . . . The pollen, when absorbed by the stigma, is conveyed down the style to the ovary, where it falls upon and fertilizes the ovules or incipient seeds. Nothing can be more beautiful or more ingenious than the mechanism by which this process is effected." Orchids especially are "still more varied and fantastic" in their reproductive mechanisms around the world. In her pre-Darwinian world, Jane underlined the ties between the human and botanical: "We know that we ourselves are 'fearfully and wonderfully made,' but how few of us are aware that every flower we crush beneath our feet, or gather only to destroy, displays as much of the Divine care and wisdom in its construction, as the frame of the mightiest giant!"

(facing page) Man orchis, bee orchis, spider
orchis, fly orchis, and lady's slipper in
Loudon's *British Wild Flowers* (1844).

1 *Man Orchis* 2 *Bee Orchis* 3 *Spider Orchis* 4 *Fly Orchis*
5 *Lady's Slipper*

Still, in addition to developing women's scientific skills, enhancing their aesthetic expertise also mattered for Loudon as a Victorian woman of taste. In *The Ladies' Magazine of Gardening*, which she edited from 1841 to 1842, she reviewed the color plates of orchids appearing in almost every issue of several rival magazines, commenting upon the exciting botany of new flowers while also grading the inherent beauty of each orchid and their aesthetic execution as prints. Loudon made the most of a tight budget in her gender-specific magazine—she had nowhere near the means to publish the lush color plates that, say, *Paxton's Magazine of Botany* was known for. Joseph Paxton had enjoyed the patronage of one of the wealthiest men in England—William Cavendish, sixth duke of Devonshire—since the 1820s, and didn't have to worry about his publication going over budget.

Loudon's 1844 book *British Wild Flowers* pushes further in equipping her audience (still a largely female one) with the inspiration and the tools to romp about the countryside and correctly identify its flora. Native English orchids here have a recurring role as important objects for scientific study, items of particular beauty, and flowers to simply appreciate and have fun with. For example, *Epipactis rubra*, the purple helleborine, is "one of the handsomest, at the same time one of the rarest in the genus." She found the genus orchis one of the most strange: "there is perhaps none in which the frolics of nature have been more fully displayed . . . [with] flowers of the most grotesque forms." The still more curious flowers of the great brown-winged orchis (*Orchis fusca*) looked like "a number of fancifully dressed up dolls hung about the spike." And ultimately she found green man orchis (*Aceras anthropophora*) utterly ludicrous, looking like "a number of little men dressed in yellow, with green caps on."

Perhaps unsurprisingly, Loudon's book that contained the most information on orchids was also one of her longest-running bestsellers—*Ladies' Companion to the Flower Garden* sold more than twenty-five thousand copies in its first years. A book designed for lit-

Early purple orchis, marsh orchis,
monkey orchis, and lizard orchis in
Loudon's *British Wild Flowers* (1844).

erate women of limited means, it offered descriptions of and growing guidance for hundreds of ornamental plants that were cultivated in gardens, arboretums, and greenhouses. Importantly, it also included the correct pronunciation of scientific Latin names, so that her still-timid audience could feel confident conversing about their plants with learned men.

Because this book had a limited budget and no illustrations, she relied on descriptive skill to paint orchids for her audience. For example, the helmut flower, *Coryanthes macrantha* "has a most singular red and yellow flower, part of which resembles a skeleton's head with the vertebrae of the neck, and part two folded bat's wings." She found laelias "extremely beautiful," but in her later years, her typical dry wit let fly as she described that lycaste "varies considerably in the different species, some of the kinds being extremely ornamental, and others very much the reverse." We can tell Loudon's work resonated with her audience because by 1850 she received a copious number of packages on her doorstep contain-

Laelia autumnalis

Laelia autumnalis. Paxton's Magazine of Botany (1839).

ing dead plants, their senders begging the knowledgeable Mrs. Loudon
to identify the brown heap of plant matter and instruct them as to why
it died. Often, her callers would not even describe whether the plant
was found in the wild or cultivated in their gardens and hothouses.

In her most significant bestseller, we also see Loudon's recurring
focus on dendrobium orchids. While she had often spent more page
space on the genus than other orchids, here she emphasized dendrobiums as

> Splendid orchideous epiphytes, which may be grown on the
> branches of trees, or in pots suspended from the rafters of the
> damp stove. They are generally propagated by taking off a joint
> of the pseudo-bulb or stem, and planting it in turfy loam, well
> drained. No water should be given till the plant begins to shoot
> from below; but in a short time, the green tips of its roots will be
> seen protruding through the loose soil in the pot, and hanging
> down over the rim.

Careful to specify many dendrobiums' specific watering regimens, she
was also attentive to the display of beautiful flowers. She suggested that
dendrobium "pots should be hung obliquely, so that the flowers may
hang down in long spikes, when they will have a splendid appearance."
Among the favorites she had grown or seen at horticultural society
meetings and frequent flower fetes were *Dendrobium nobile*, *devoni-
anum*, *discolor*, *moniliforme*, *fimbriatum*, and *macrophyllum*. Loudon
may have emphasized *D. nobile* in her books because her mentor John
Lindley described it first for science. She found that the most ornamental
species were natives of Nepal and other areas of the East Indies, which
would include *D. nobile* and most of the list above. Here Loudon reveals
something about her own temperament as well as her lifelong appreciation of all things that were hardy, adaptable, easily grown, floriferous,
and graceful in their season.

Loudon commented of *Dendrobium macrophyllum* published in *Paxton's Magazine of Botany* (1841), "very beautiful . . . a most abundant flowerer, and very valuable plant."

At the end of *Ladies' Companion*, the author takes a moment to wax about a love for plants and scientific botany as a requirement of being a good gardener: "No one, in fact, can ever make a good gardener, who has not a sincere love for plants." Invoking the seventeenth-century English parson-naturalist John Ray, she concludes that "love was the inventor, and is still the maintainer of every noble science . . . it is that which animates and renders [plants] strong and vigorous; without which they will languish and decay." Love, beauty, science, and education—Loudon's many messages had become fused in orchid growing for a wide audience of flower-mad Victorians.

Loudon died at the age of 51, overworked and penniless in her attempts to pay off her husband's debts. During her short life, Loudon's publications celebrated floriculture as an energetic activity, one inclusive of hard work, aesthetic sense, and intellectual effort. In no small way, it laid the groundwork for the labor and scientific brainpower women wished to be known for in other areas of their lives, too. Jane's orchid publications provided an inviting path for many women to take their first steps into wider natural and societal worlds.

DENDROBIUM NOBILE

Of the many species one may choose to grow in honor of Loudon, an orchid with the appropriate mix of delicacy and resilience is *Dendrobium nobile*. In style in Victorian-era greenhouses and fashion, its flowers come in rose, mauve, and cream tones, with a ruffled lip, much like the train of a smart dress. Sweetly scented but with a robust cane, this orchid is a mix of grace and strength.

Paul Atkinson/Shutterstock.com.

Orchid Details

— Place of origin: India, Burma, Thailand, Vietnam, China
— Blooming season: January to May, peaking in March
— Flowers last for: one to three months
— Plant size: height to 2 feet tall; undivided plants grown to specimen size can be several feet wide
— Flower size: 3 inches wide, several flowers on each inflorescence
— Fragrance: varies by time of day. Usually honey-sweet in the day and hay-like at night. Can also take on floral or musky tones.
— Plant habit: sympodial with upright bunching canes

Orchid Needs

— Light: medium to high; eastern or southern window or
under lights
— Temperature range: summer days 70°F–85°F, 45°F–58°F
nights year-round; can be grown outside in summer and
left out until just before a first autumn freeze; likes to be
kept at a relatively constant 55°F as flowers form
— Humidity: ideally 80% in summer, 60% in winter
— Water: heaviest water May through October, up to 11 inches
per month; very little to no water November through April
— Growing medium: fine to medium orchid bark; likes to be
pot-bound
— Fertilizing schedule: half- to full-strength balanced fertilizer
in spring and summer when plants are actively growing
— Seasonal changes: two- to three-month dormant dry rest
in fall; leaves will drop off of canes, but continue to give it
bright light
— Special requirements: This orchid must have a dry, cool
winter rest to bloom. Reduce water in fall to almost nothing; water only to prevent canes from shriveling. Resume
watering only when flower buds on canes are well-formed;
watering early usually induces keikis (plant babies) to grow
instead of flowers.

Chapter 10

S

Orchids and Steel

We might set 1907 as the high-water mark for the Gilded Age orchid obsession. On Sunday, February 17 of that year, the *New York Times*—amid a political cartoon of Teddy Roosevelt proclaiming "Let Us Have Peace" while throwing out "Big Sticks," an article about the "men who handle Mr. Rockefeller's $43,000,000 fund," and tell-all details about the creation of an elite women's "Colony Club" to which Mrs. John Jacob Astor belonged—declared in large font with inset images of paphiopedilum, cattleya, and laelia orchids that "The Orchid Craze Is At Its Height In Fashionable New York."

Of course, *the* mark of one's arrival into high society was to regularly spend stupendous amounts of money on orchids. The Astors, Rockefellers, and Carnegies bedecked their opulent homes with rare and beautiful cut flowers. The extended Vanderbilt family set a trend that a lavish wedding for one's daughter must include glittering white orchids. Every facet of a wealthy couple's nuptial decorations became a weekly feature in national newspapers. Money spent on perishable beauty was still money well spent, as it served to advertise the family business.

Several of the wealthiest Americans did more than flaunt their wealth with orchids. Some enhanced public and private orchid appreci-

Part of Helen Gould's orchid greenhouse (late Jay Gould estate), 1894.

ation, including steel and real estate magnate Henry Phipps, who made significant donations to regional botanical gardens and public greenhouse construction. Businessmen Henry Clay Frick, Marshall Field, H. J. Heinz, C. L. Tiffany, Hamilton Twombly, and Cornelius Vanderbilt were all known for their large private conservatories, employing dozens of gardeners and stocked with rare plants. Some millionaires additionally instructed their orchid growers to submit their coveted flowers in regional horticultural society shows, and took home copious awards.

Many well-known tycoons were orchid collectors, but in truth knew little about orchids. In this category fit railroad mogul and speculator Jay Gould, renowned for his cunning business sense—except when it came to orchids. The *New York Herald* wrote in 1896 that "many exquisite and rare varieties were numbered [in his collection], but an excessive price was paid for the most of them . . . careful, conservative buying by an orchid expert would have gathered it together at almost a fraction of the sum actually paid out."

Helen Gould, daughter of Jay Gould, spent her life simultaneously honoring her father's love of orchids and unraveling his reputation as

Helen Gould (c. 1910).

the dark king of all robber barons. In contrast to Jay's ruthless railroad empire, manipulation of the stock market, bankrupting of his business partners, and successfully hoarding enough gold to set off a nation-wide panic, his daughter cut a figure of philanthropy and a different kind of orchid appreciation. She was a patron of the arts, and donated to libraries, churches, and YMCAs across the East Coast, while paying special attention to the needs of homeless families, disabled children, and soldiers during and after the Spanish American War. Privately, she adopted three children and fostered a fourth, and became the matri-arch of an extended family that was otherwise riddled with scandal. The uproar induced her to also raise her siblings' children.

Four years after her father's death, when Helen was twenty-eight years old and a recent graduate of NYU Law School for Women, New York's *The Journal* interviewed the heiress about her orchids, "one of the finest collections in the country." The journalist wrote of the contrast between orchids' general usage and Gould's refined floral connoisseurship. The former had explicit rules for utilizing the right orchid at the right func-tion: modest gray cypripediums with transparent petals for debutantes; during Lent, lavender cattleyas were in order for dinner parties; and for weddings, white dendrobiums and phalaenopsis were a must. And for the very fashionable, "the violet-colored cattleya is most sumptuous in effect, worn with black velvet and chinchilla, and the gray and yellow orchids are associated on the corsage of a mouse-color or brown tailor suit." All arrangements, from the boutonniere to the wedding bouquet, included many sprigs of ferns for the "effect is tropical in the extreme," but it was thought rather gauche to mix orchids with other flowers.

Better by far than all the previous guidelines and social graces, *The Journal* intoned, was Gould's passion for orchids; her attention meant something more than collecting beauty or parading one's wealth in botanical form. The heiress's dedication to cultivating orchids came with a promise to allow the public, especially children, to access her conservatories to admire and learn from the plants. Gould was a con-scientious lay botanist, an active supervisor of her hothouses, and had

Charles G. Roebling (c. 1915).

the spirit of an informed collector. Helen did not fall for hothouse fads. Indeed not. Rather she was as "proud of adding a new orchid to her already wonderful assortment as most women would be of a new diamond ring or jeweled girdle [belt]."

Fewer still were millionaires at the turn of the twentieth century who developed a genuine passion for hybridizing orchids and participating in the international orchid trade. This love of orchids, combined with the endless funds to enact luxuriant dreams, is something that can be traced through the historical record and within the genes that popular hybrid orchids still carry today. A rarer breed of industrialists became passionate about orchid innovation, building international orchid networks, sharing their knowledge with the public, and donating new hybrids to charitable auctions. One of these was Charles G. Roebling (1849–1918). He was internationally respected in the world of orchids and an even more exacting enthusiast than Helen Gould.

As the president of John A. Roebling's Sons Company, Charles ran a profitable engineering firm that oversaw the completion of the Brooklyn Bridge. Less well known are his later ventures building steel and wire mills that produced materials for the Williamsburg Bridge and

Great East River Suspension Bridge, now known as the Brooklyn Bridge.
Currier and Ives (1857–1907).

the Golden Gate Bridge. He additionally manufactured wire rope for
the front during World War I, and rigging for the Panama Canal, air-
planes, trams, mines, elevators, and many other structures. At their
peak, the factories and machines he designed produced thousands of
miles of wire per day. And when the Roebling brothers needed a more
reliable source of steel, Charles designed and built their company town
of Roebling, New Jersey, near their factory so that production would
run smoothly. He had long lived in the shadow of his father, John Roebling, who died painfully of tetanus in 1869 after sustaining an injury
during surveying he did for the construction of the Brooklyn Bridge.
The elder builder was full of bombast and couldn't keep from boasting that the elegant structure would "not only be the greatest bridge in
existence, but . . . the greatest engineering work on the continent, and
of the age . . . a national monument and a work of art."

But Charles possessed a gardener's quieter soul. Having fallen in love with orchids as a young man, throughout the late nineteenth century he financed multiple botanical missions into the mountains of South America and throughout the Amazon basin, as well as to Samoa. Most importantly, he contributed to a loose international coalition of hybridizers in what has been called the First Golden Age of paphiopedilum breeding. Not only were new species discovered, but breeders also shared pollinia to create primary hybrids and more complex crosses.

Bridge building and orchid breeding became entwined metaphors for Roebling, who cultivated both plants and cities in his lifetime. Abram Hewitt, future mayor of New York City and speaker at the Brooklyn Bridge Opening Ceremonies in 1883, aptly described the bridge as "not merely a creation," but "a growth." Hybrid vigor—the increased size and strength developed in Roebling's crossbred orchids—was also found in the mixture of iron and carbon in Roebling's bridges: steel offered the required sturdiness and flexibility for a great arch.

Perhaps an extension of his professional thought process, in the early 1880s Charles began to collect orchids with alacrity. He hired two of the most respected orchid growers of the age—Henry T. Clinkaberry and Jason Goodier—to take care of his greenhouses as well as hybridize several genera in his burgeoning collection. Roebling's orchids, like all orchids, were hybridized by collecting the pollinia of one plant with a toothpick and transferring it to the stigma of another. Once the orchid is cross-pollinated, the flower falls off and the seed pod grows. Paphiopedilum seed pods, for instance, take nine to eighteen months to mature. In Roebling's day, when ripe, millions of dust-like seeds were broadcast on moss or at the base of mother plants. Germination was very low, and if lucky, he might find seedlings growing after another one or two years. Once stout enough and separated into their own pots, paphiopedilum plants would take another three to five years to bloom. Whereas tolumnias can be grown from seed to flower in as little as two years, cattleyas take an average of eight years to bloom. All orchid

Paphiopedilum Niobe, a primary hybrid of P. fairrieanum and
P. spicerianum, was registered by British orchid breeders Veitch & Sons in 1889.
Roebling soon used the cross to make other paphiopedilum hybrids. Warner,
Williams, and Moore, Orchid Album (1893).

hybridizers are patient in this way, and Roebling was willing to put in the time to create beautiful new flowers. Roebling also began to contribute horticulturally to his community in other ways: he bedecked his church in orchids on holidays, and sponsored city greening efforts.

Roebling's hybridizing program for his favorite orchid genera, cattleyas and paphiopedilums (called cypripediums in his day) required the same sort of long-term planning, development, and adaptation to adverse conditions that his bridges did. Paphiopedilums are perhaps both the most anthropomorphized and oddest orchids for their often hairy and alluring petals and pouches. In addition, paphiopedilums seem as though their heads (the flowers) are suspended in air, too heavy for their necks (thin stems) to support. Much like engineering a suspension bridge's perfectly balanced "catenary curve," engineering attractive curves and proportions in orchids—along with a strength to withstand inhospitable environments—marks successful orchid hybrids. And in the long term, it is not just the creation and naming of a new orchid that matters in the wider world of horticulture; it is whether the hybrid has enough positive traits to be of use in future orchid breeding. By this measure, Charles Roebling was successful in revolutionizing steel wire construction and in adding to the growing diversity of beautiful, collectible orchids.

Roebling was not a saint, of course. People remarked about his gruff personality, forged while breaking regular strikes at his factories and rebuilding after dozens of devastating fires, which occurred by accident and by arson (disgruntled employees and anarchists were regularly accused of destruction of property). His factories spilled wastes into the rivers they were built on, and tons of slag exited their furnaces. And although close with his immediate family, including his brother Washington Roebling, wife, daughters, and a single son—of at least three born—who lived beyond early childhood, once the Roebling's Sons Company was incorporated, the brothers often went six months without speaking to one another.

By 1895, Roebling had one of the finest orchid collections in the

Vol. VI.   CHICAGO, DECEMBER 1, 1897.   No. 120.

Award-winning basket
of Roebling's orchids,
created by Clinkaberry,
on the cover of
Gardening magazine
(1897).

United States, and cultivated the flowers in a sprawling complex of five greenhouses at his home in Trenton, New Jersey. His cool house sheltered thousands of masdevallias and odontoglossums. His temperate houses were awash in epidendrums, dendrobiums, and cymbidiums, coelogynes, lycastes, and zygopetalums. And his warm houses grew vandas, calanthes, and appropriate-temperature paphiopedilums. He attested that "many hundreds" of species and primary hybrids were contained in his paphiopedilum collection. His best orchids were used as breeding stock to invent or re-create award-winning hybrids. His collection included *Paphiopedilum insigne* in brown and yellow tones, *Paph. charlesworthii* with its enviable pink dorsal sepal and round flower, the twisting spotted petals of *P. Morganiae, P. leeanum*'s green-tinged dorsal sepal, the golden-striped *P. praestans,* and the hairy, undulating, magenta petals of *P. Niobe.* Roebling's cattleya house alone was sixty feet long; there, he focused on white flowers, including alba forms of *Cattleya aclandiae, gaskelliana, mossiae,* and *trianae.* He was sure to spend Sundays and rare holidays in his orchid greenhouses,

Paphiopedilum insigne as *Cypripedium insigne* in Warner, Williams, and Moore, *Orchid Album* (1893).

he and his gardeners racking up awards at regional orchid shows for almost forty years.

In March of 1904, *American Gardening* ran a first-page feature on one of Roebling's most valuable plants: a division of *Paphiopedilum insigne* 'Sanderae,' described as a "chaste and rare variety," as well as the world's "finest known specimen of this rare orchid." From the Latin, *insigne* means eminent and distinguished—and these terms are certainly accurate if we also apply them to the orchid's role in paphiopedilum breeding for more than 150 years. *Paphiopedilum insigne* has been used as a parent in nearly three hundred hybrids and lays claim to ancestry in more than nineteen thousand other progeny. It is one of the four most important paphiopedilum breeding species, the other three being *P. spicerianum*, *villosum*, and *bellatulum*.

Paphiopedilum insigne claimed its place in the halls of modern orchid hybridization through its stalwart constitution, known to with-

Cattleya C. G. Roebling (1895) 'Blue Indigo.'

stand all kinds of abuse on Victorian windowsills. Its flowers sport an array of autumn colors and attractive spots, and its incurved petals were thought to epitomize modesty, making the orchid appropriate for all occasions. It remains a dependable species, producing multiple new fans and flowers annually.

Avid members of the Society of American Florists and Ornamental Horticulturists, Roebling and Clinkaberry together and separately won dozens of First Class Certificates for their orchids—the highest possible flower quality award—from the New York, Massachusetts, and Pennsylvania Horticultural Societies. They additionally mounted large exhibits for the American Institute in New York. While regularly lauded for their individual plants, the artistry of their displays sometimes left something to be desired. At the Massachusetts Horticultural Society Orchid Show in May of 1910, a critic wrote that "it was impossible for anyone to see all the orchids without going through the maze of winding paths with many plants hanging overhead. In fact, it was an exhibit that could not be seen." Still, Roebling's display contained the most variety in orchids, and took home second prize.

More importantly, Clinkaberry and Roebling also registered no less than twenty new cattleya, paphiopedilum, laeliocattleya, and zygolum hybrids. Sadly, records of the genetic trees of most of them have been lost to history or were not properly recorded. Yet a handful of their orchids have a continuing place in breeding today.

Roebling's major contributions to the orchid world underlined his relationships and priorities as an orchid breeder. As early as 1895, Roebling was honored by Henry Frederick Conrad Sander, "Orchid King" and the preeminent orchidologist in Europe. Sander named a primary hybrid of two cattleya species, *C. gaskelliana* and *C. purpurata*, after the businessman. *Cattleya* C. G. Roebling (1895) is a large white orchid with a royal purple lip. Roebling registered a few hybrids of his own in the meantime, and by 1903 donated a flamboyant new orchid to a Royal Horticultural Society fundraiser organized by Sander. He couldn't resist naming the award-winning orchid after himself: *Zygopetalum*

Roebling's *Paphiopedilum* (then *Cypripedium*) Garret A. Hobart on Lenox china plate he commissioned (1906). *Private Collection. Photo © Christie's Images/ Bridgeman Images.*

Roeblingianum (today *Zygolum* Roeblingianum), a bloom with chartreuse and maroon petals, a purple column, and veined pink lip. With the 50 British guineas (about $7,300 USD today) the new hybrid fetched, the RHS was able to pay for a significant portion of the construction of a new Horticultural Hall. Charles also went on to name after himself the last orchid he would hybridize: it was another primary hybrid, *Cattleya* C. G. Roebling (1916), a cross of *C. harrisoniana* and *C. mendelii.* It too is a large white orchid, but it carries a ruffled, light lavender lip with lavender splashes on its petals.

Roebling often commemorated US presidents and vice presidents with his hybrids, participating in a trend that has now extended to naming orchids after First Ladies. Roebling dedicated many of his lady's slipper orchids to White House leadership through time: *Paphiopedilum* Adamsii, James Garfield, Abraham Lincoln, Franklin Pierce, and James K. Polk; his most significant accomplishment in the genus, ironically enough, was *Paphiopedilum* Garret A. Hobart, named after the vice president to William McKinley who died after two years in office in 1899. The flower is comprised of three of the nineteenth century's most popular species (one-half *P. insigne*, one-quarter *P. villo-*

sum, and one-quarter *P. spicerianum*). Hobart's strong constitution, fall colors, and ruffled petals have led it to appear in the ancestry of sixty-seven paphiopedilum progeny.

To offer some perspective on a century's worth of advancements in orchid hybridizing, compare *Paph. insigne* to *P.* Stone Crazy, hybridized in 2009, whose lineage is almost half *P. insigne*. Contemporary international orchid judging associations award the most points to perfectly round, large flowers: that aesthetic has been the basis of the awards system for many orchid genera for generations. Hybridizers seek to create award-winning and popular flowers, and thus emphasize certain orchid elements. So, whatever your taste, be assured that there is a paphiopedilum with colors, stripes, spots, warts, and/or hair that will appeal to you.

For a man who ran a sharp business, Roebling never skimped on orchids—he spent several hundred thousand dollars on his collection and breeding program over the course of his life. Roebling wrote in 1901 that "one of the greatest pleasures of orchid culture is derived from hybridizing. It seems a long time to wait several years for a seedling to bloom, but when the hour comes that gives us a variety that not only is beautiful, but has never flowered before, it repays us for all the trouble

Paphiopedilum Stone Crazy.

that has been given to the new plant." Upon his death in 1918, he was still cultivating more than seven hundred species, varieties, and hybrids, with several thousand additional flowered and unflowered seedlings in his greenhouses. Charles's heirs sold the collection within months of his passing for $28,000 ($450,000 today); it was undervalued, due to the economic effects of World War I and the Spanish Flu. Clinkaberry and Goodier went to work for other wealthy orchid enthusiasts.

Although he rarely publicly flaunted his wealth, Roebling had immersed himself in orchidomania from 1906 to 1912, when he paid $3,000 (about $85,000 today) for a commission to paint thirty-two Lenox porcelain plates with images of orchids. The flowers were reproductions of blooming specimens, drawn in situ in his Trenton greenhouses. His plates, with elaborate gold-etched borders that included his initials, showcased both his named hybrids and orchid species. Included were star-shaped *Epicattleya* Nebo, by-then classic purple-webbed *Vanda coerulea*, hot pink *Cattleya* Mantinii, leopard-spotted *Odontoglossum triumphans* (now *Oncidium spectatissimum*), and several other assorted genera. And as any true orchid aficionado would require, the Latin names of the species were printed on the plates' reverse—also in gold. Roebling may very well have continued to commission Lenox plates painted with his favorite orchids, but in 1912 the artistry came to an abrupt halt, and Roebling's participation in orchid society shows declined. Less than two weeks after the last two orchid plates were painted (*Oncidium* Rolfeae and *Masdevallia veitchiana* var. grandiflora), Charles's 31-year-old sole son, Washington Roebling II, went down with the *Titanic*. Charles was shattered and increasingly reclusive for the rest of his life, dying six years later.

So it seems that the best and the worst of the Gilded Age and Progressive Era trends were represented in full in the orchid community. It was an age of excess, outrageous inequality, expansion, and innovation. The orchid was "the flower of the moment, expensive and fragile," wrote *The Journal* in 1896. Expensive, certainly. Fragile? Not at all.

PAPHIOPEDILUM INSIGNE

Many growers find that next to moth orchids, slipper orchids are the easiest to grow in the home environment. Yet, as terrestrial orchids, they require care that is somewhere between a standard houseplant and an epiphytic orchid.

Orchid Details

~ Place of origin: Northeast India and Bangladesh

~ Blooming season: fall and winter

~ Flowers last for: one month or more

~ Plant size: at full size, the plant can be 1.5 feet wide, flowers held 12 inches high above the foliage

~ Flower size: 3-4 inches across

~ Fragrance: none

~ Plant habit: strappy green leaves, flowers sit well above the foliage; can grow into a specimen the size of a small bush

Orchid Needs

— Light: medium-low, shaded
— Temperature range: winter: as low as 40°F–50°F nights, 65–75°F days; summer: 55°F–65°F nights, 76°F–80°F days
— Humidity: 65%–80% ideal
— Water: water often in warmer growing months, much less in cooler, slow-growing months; do not let fully dry in warm months
— Growing medium: small to medium fir bark and perlite; add chopped sphagnum if humidity is lower than ideal; repot often—at least annually
— Fertilizing schedule: use a half-strength balanced orchid fertilizer throughout the year, and flush regularly with pure water
— Seasonal changes: needs a four-month cooler, drier winter to bloom well
— Special requirements: cool winters, otherwise good grower under lights, on windowsills, in greenhouses, or outdoors if the climate allows

PART IV

Orchid Culture

Chapter 11

The Flowers, Fashion, and Friendships of Empress Eugénie

I magine a girl raised on the whimsy of Washington Irving, one who also grew up under the influence of Sir Walter Scott. A girl as willful as the horses she rode, as passionate as the early nineteenth-century gothic romances she swallowed whole. She crafted the story of her birth: "I was born during an earthquake . . . My mother's *accouchement* took place beneath a tent in our garden. What would the ancients have thought of such an omen? Surely they would have said I was destined to unsettle the world." This was Eugenia, daughter of Spanish nobleman Cipriano de Palafox, eighth Count of Montijo, and Scotch-Spanish mother María Manuela Kirkpatrick. Eugenia's riotous imag-

Eugénie, Impératrice des Français, wearing a flowered lace dress for her portrait. Léon Alphonse Noël after Franz Xaver Winterhalter (1854).

ination became real life in 1853 when she married Napoléon III and was crowned *Eugénie, Impératrice des Français.* For Empress Eugénie (1826–1920), orchids were part of both the everyday romance and elite royal trendsetting she conjured in Second Empire France and through-out the Western world.

It is true that Washington Irving lived near and educated Eugenia and her sister for a long summer in 1829. While still basking in the fame he first garnered a decade before for "Rip Van Winkle" and "The Legend of Sleepy Hollow," Irving was allowed to reside in the ruins of the ancient palace-fortress in Grenada, Spain, while he wrote *Tales of the Alhambra.* As warm-up to the fanciful myths and nonfiction he collected there, Irving described his introduction to the Alhambra by a "good fairy who was to conduct us through the palace. Under her guidance we crossed the threshold, and were at once transported, as if by magic wand, into other times and an oriental realm." He waxed at length about the Arabic poetry inscribed on the palace walls, the simple beauty of Spanish women wearing flowers in their hair, and the

Alhambra palace-fortress in Grenada, Spain. *Marques/Shutterstock.com.*

verdure outside his window. "How beauteous is this garden, where the flowers of the earth vie with the stars of the heaven!" By the age of three, Eugenia was instructed by the master himself in a deep love of gardens, fairytales, distraught heroines, and noble deeds. Eugénie the empress, throughout her adult life, returned to nature-based fantasy in times of stress and celebration.

Among many other surprising family connections, Eugenia's father fought against his countrymen as part of Joseph Bonaparte's army in the Peninsular War, lost an eye, and was honored in Paris by Emperor Napoléon I. The future empress was close with her mother, and her elder sister María Francisca (nicknamed Paca) was an affectionate confidante—although in her mid-teens Eugenia swooned over two prominent men who loved Paca more. The family member most cherished by Eugenia, however, was her father. He died when she was only thirteen, and the future Eugénie became hopelessly devoted to the Napoleonic grandeur he represented. By her late teens, she developed a deep passion for deep purple violets, symbolically Bonapartist flowers. When spring turned to summer and they became hard to find, she employed a shepherd to bring violets down from the Sierra Nevada.

In the early 1850s, the French were understandably divided about President Charles-Louis Napoléon Bonaparte III's sudden and violent transition to emperor, reanimating a family will to power that they thought had died a generation before. His coup in December 1851 was seen in turns as a bloody and egregious abuse of power and the best option for defending universal male suffrage. Although widely controversial, the coup was then approved through national referendum, and after a second vote, Napoléon III was declared emperor for life.

Soon known as the "Sphinx of the Tuileries" and notorious ladies' man, Charles-Louis was suddenly in need of a wife to produce a legitimate heir. He approached the royal houses of Sweden and England, but they demurred because of France's political instability and his Catholicism. Although Eugenia was not flush with royal connections, she was Catholic, and guaranteed passion for the Napoleonic cause. When they wed, Napoléon was forty-four years old; newly crowned Eugénie was almost two decades his junior. She was also the very definition of Victorian beauty for her pale face, strawberry-blonde hair, violet eyes, and perhaps most of all, her shoulders—displayed nicely in the décolletage-bearing fashions of the day. (Her beauty did not slow Napoléon's dalliances with other women, however.)

There is still debate about whether Eugénie was merely an "ornament of the throne" or a powerful woman pulling political strings behind the scenes. There is no doubt, however, that she was the nineteenth century's foremost fashion maven, a major patron of the arts, and an advocate for the education and welfare of women and children when only 45% of the French female population could read and write. She was additionally outspoken for improved hospitals, schools, and prisons throughout her reign.

As much as she loved the elegant trappings of being an empress, the social reformer sometimes took precedence. As just one indication, upon her betrothal to Napoléon III, the City of Paris presented her with 600,000 francs earmarked for engagement jewelry. In one of her first acts as empress, she decided to use the money to, instead, build a boarding school for poor French girls—in the shape of a two-tiered pendant necklace. By many counts, she was among the most powerful women of the nineteenth century; more powerful, perhaps, than even Queen Victoria, because as constitutional monarch a queen did not dictate national policy.

It turned out that the two monarchs became fast friends—their most important meetings in life punctuated with orchids, fine gardens, and haute fashion. The sovereigns had long appreciated each other from afar. Upon first seeing Eugénie, Victoria wrote that she was "beautiful, clever, very coquette, passionate and wild." Their earliest extended meeting, when Eugénie was twenty-nine years old and Victoria thirty-six, was awash with royal pomp and garden appreciation. Queen Victoria and Prince Albert welcomed the imperial couple to London in April

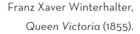

Franz Xaver Winterhalter,
Queen Victoria (1855).

1855; they attended the opera, toured the palace, and the four appeared at ceremonies in their honor. A key event of the several-day stay was a formal reception at the Crystal Palace, which had recently been relocated from Hyde Park (its original home, built for the international 1851 Great Exhibition) to Sydenham Hill. As *The Scotsman* and other newspapers reported that spring and summer, a visit to the Crystal Palace "both in quality and extent . . . exceeded the most sanguine expectations. There was not only an unrivaled show of old plants, but a great display of new and rare specimens, and the admiration of the orchids especially was universal and unanimous."

The couples were greeted by thousands of well-wishers. The queen and the crowds who saw Eugénie that day commented on her grace and her exquisite lilac dress. The royals toured the building displays and its wider grounds, and Eugénie thrilled at the Crystal Palace's replica of her childhood haunt, the Alhambra's Court of Lions. They chatted at length about the flowers, shrubs, and landscape design. And they would have encountered several orchids, as the Crystal Palace displayed exotic plants without stop.

That season, the magnificent glass house held a large specimen of the crystalline and romantically arching *Coelogyne cristata*, among several other orchid genera. In spring and summer, it additionally showcased orchids from China such as the deliciously scented *Aerides odorata* and *multiflora*, frilly pastels of *Dendrobium nobile* and *moniliforme*, stunning pinks of *Papilionanthe teres*, and daffodil-shaped *Phaius tankervilleae* and *albus*. Central and South American orchids on display were the lemon-yellow arcs of *Oncidium sphacelatum*, the coconut cream pie orchid *Maxillaria tenuifolia*, large and lovely *Cattleya purpurata*, and the classic *Cattleya mossiae*, *intermedia*, and *aclandiae*. Orchids from Thailand and Malaysia included purple-pouched *Paphiopedilum barbatum* and *Phalaenopsis amabilis*, the white moth orchid from which most large phalaenopsis hybrids stem.

As the state visit came to a close, Napoléon III and Eugénie immediately returned the invitation for Victoria and Albert to visit them that

COELOGYNE CRISTATA Lindl.
Uitgave van J.B. Wolters, Groningen

Coelogyne cristata. Nederland's Plantentuin (1866).

summer and attend Paris's Exposition Universelle. Within weeks, the
date was set for a ten-day visit in August.

If anything, the state visit to Paris was even more lavish and awash
in flowers than the London affair. In addition to Paris's normal summer grandeur—its parks exploding with flowers—the Tuileries' greenhouses were packed with Eugénie's favorite orchids from China, India,
and Mexico. As Émile De Puydt's lushly illustrated book *Les Orchidées*
detailed, orchid collecting had grown rapidly in France starting in the
1830s. He waxed, "orchids are the life and the charm of forests, jungles,
mountains . . . they lavish their flowers there, with incomparable freshness and grace, and their scents of a strange sweetness on the stars." He
offered a beginner's list of orchids for his readers to grow, including

Franz Xaver Winterhalter, *Empress Eugénie Surrounded by her Ladies in Waiting* (1855). Eugénie sits left of center in white gown with purple ribbon. *Chateau de Compiegne, Oise, France/Bridgeman Images.*

cattleya, dendrobium, cypripedium, oncidium, and several other genera. To whet their excitement, he promised that there were more than one thousand species available to the interested grower, emphasizing *Coelogyne cristata*, a classic and elegant orchid, reliably bearing large flowers and easy to grow in a cool greenhouse.

Napoléon accompanied Victoria and Albert to the exposition, where they viewed exhibits from dozens of countries. Although officially presiding over the Paris Expo, Eugénie did not attend that day, likely because she was two months pregnant, not feeling well, and feared another miscarriage after two previous losses. Victoria, Albert, and Napoléon encountered Eugénie's likeness at the exposition, however. Court painter Franz Xaver Winterhalter's recent portrait of the empress alone, as well as a life-sized painting of her in a group portrait, *Empress Eugénie Surrounded by her Ladies-in-Waiting* (1855), were prominently displayed in the exposition's *Salon d'honneur*. The sumptuous image of the latter presented women in clouds of taffeta and tulle, wearing garden flowers of late spring: lilacs, violets, iris, honeysuckle, hellebore, roses, and camellias. The painting popularized a hairstyle named after Eugénie—the entire party's tresses are dressed *à l'imperatrice*, parted down the middle, elaborate curls set off the face, with flowers and ribbons twisted throughout. Eugénie sits atop and left of center on a hummock in a dress of white silk and tulle decorated with violet ribbon. She is a fairy queen in her woodland garden, embodying the fertility of nature. In choosing Winterhalter as her court painter, Eugénie was keen that he represent both the fashion of her ideal empire and its bucolic whimsy.

The painting became a celebration, a fashion statement, and a political statement. Nine ladies elaborately arrayed and at leisure in a sylvan garden—presented much like gorgeous flowers themselves—highlighted the dazzle of imperial splendor that the Second Empire was already well known for. Empress Eugénie's fashion prowess had been at first jeered; the empire was nicknamed "*la reine Crinoline*," attempting to belittle its impact as a simple lady's dress support. And in truth, by 1856, Eugénie had popularized the cage-crinoline, in every way an improvement over petticoats. Crinoline used hoops of steel wire to give a wide dress its shape, instead of dozens of layers of heavy fabric. A lover of horses and the outdoors, the sporting empress stated plainly that she liked crinoline because it allowed her to use and move her legs easily. And a few years later, she requested that her designer create a "walking crinoline" four inches shorter, so that she might enjoy her garden and delight in country walks without getting muddy. Throughout the period, Eugénie-inspired "crinolinomania" swept France, Britain, and the United States. In addition to her efforts as social reformer at home, she had given style to the mounting international pressure for women's emancipation.

On the royal couple's penultimate evening in France, Eugénie staged one of the most magnificent balls of her career as empress. Hosted in Versailles' Hall of Mirrors, she called for costume dress on an eighteenth-century Louis XV theme. To begin the festivities, Eugénie set herself at the top of a marble staircase covered with royal purple carpet, its balustrades dripping with masses of white orchids, ferns, and mosses. Eugénie sparkled like a jewel, mostly because she was encrusted with so many: she wore a white dress embroidered with diamonds, blades of long grass sewn into its folds and accessorized with more diamonds in her hair. Victoria swooned over the event and Eugénie's performance: the queen said that above all, the empress looked "like a fairy queen or nymph."

At her *fêtes imperiales* and in her interior design, Eugénie was inspired by, and certainly helped drive, the mid-nineteenth century love of flowers. Orchids were a part of a constant quest for novelty and variety in both professional horticulture and the wider world of

fashion—two realms Eugénie worked hard to combine. Throughout the year, several greenhouses in Paris forced blooms for market; Belgium, as well, was a major source of flowers, especially orchids, for Parisian buyers. While Eugénie had a constant supply of fresh blossoms at her fingertips from the Tuileries' greenhouses, people of all classes soon shared her appreciation of flowers choosing among the differently priced blooms. Artificial flower businesses boomed as well. Like Eugénie, women took to wearing real or artificial flowers in their hair, on hats, and pinned or sewed to their dresses and handbags.

Empress Eugénie's diamond *Coelogyne cristata* brooch. © *Christie's Images/ Bridgeman Images.*

Because of her beauty, popularity, and possibly also her positive effect on flower sales, plant breeders for more than a century took inspiration from Eugénie. Dozens of flowers have been named after the empress, including varieties of begonia, passionflower, iris, fuchsia, azalea, rhododendron, chrysanthemum, and rose. At least two orchids were named for Eugénie in her lifetime: *Cattleya mossiae* 'Empress Eugénie' and *Odontoglossum* Empress Eugénie.

From the moment that her courtship with Napoléon had commenced, Eugénie also began to amass a collection of botanically themed jewelry. His first gift to her was a brooch of emeralds in a clover-leaf formation, set in diamond "dew." She received many spectacular floral pieces: jewelry in the shape of currant leaves and of lilac sprays, a crown of diamonds in the shape of an olive wreath, and another glittering head ornament composed of diamond wheat stalks, cornflowers, and grass. Perhaps most impressive, as the *London Daily News* reported, was a "wonderfully light brooch of brilliants made to represent an orchid with long pendant leaves and drops . . . another ornament to which pure taste gives additional value . . . There is not one [piece of jewelry] which does not suggest State pageantry and Imperial profuseness." Given which orchids were available and popular, the brooch in question is very likely a jeweler's elegant interpretation of *Coelogyne cristata*, designed as if mounted on a tree. The orchid was entered into the scientific record by Charles Lindley in 1824 and grown widely throughout England and Europe by the 1840s.

François du Buysson, author of another popular book of the period, *L'Orchidophile*, described *Coelogyne cristata*'s appeal: "it is the most beautiful species of its kind . . . its blooms are wide open, of a pure white, with a sweet odor . . . the labellum very large, three-lobed, also white-pure, but adorned in its center by five ridged or barbed stripes, the color of a bright egg yolk." He celebrated the orchid as a "magnificent" vigorous grower that flowered for more than a month at a time, offering beauty and brightness in the coldest part of the year, winter and early spring.

Coelogyne cristata has several other qualities that link it with

Eugénie and this particular moment in French couture. She was known for wearing white and pastel colors. The choice of white coelogyne and cymbidiums in her jewelry and décor points to her wish to be thought classic, timeless. (It was also a nod to Queen Victoria, who had created her own instant fashion trend in 1840—that of wearing an elaborate white wedding dress.) The empress was known for her large violet eyes, which turned down at the corners. In addition, one of the hallmarks of nineteenth-century fashion was that it emoted an air of an ideal Victorian woman's "drooping restraint." Here, Eugénie made a choice that suited her taste in fashion, if also her personality: *Coelogyne cristata* is one of the most elegantly droopy orchids around.

The love of orchids in France coincided with an explosion of gardening. As just one example of the wider appeal of gardening on the continent and in Britain, on several days of the imperial couple's 1857 trip to Victoria and Albert's summer home at Osborne on the Isle of Wight, one could find Eugénie and Victoria planting trees together after lunch. But the nineteenth-century rage for nature during a period of widespread industrialization was especially acute in Paris. At the beginning of his rule, Napoléon III directed Baron Georges-Eugène Haussmann to "renew" Paris. The two men viewed France's capital as a still-medieval city whose overcrowding and lack of sanitation had led to waves of disease. They commenced upon a twenty-year urban renewal project that within months razed fifty thousand buildings and displaced hundreds of thousands of people, tearing down much of central Paris and remaking it with wide boulevards, standardized architecture, and a new aqueduct that ensured fresh water. With input from Eugénie, Paris and surrounding areas also witnessed the creation or redesign of eight square miles of parks, gardens, and squares during Napoléon III's reign.

One of the most acclaimed writers of the age, Charles Baudelaire, described the violence of the Hausmannization of Paris and the simul-

Charles Baudelaire, *Fleurs
du Mal* cover with invented
phragmipedium orchid, 1900
(originally published 1857).

taneous decadence of the Second Empire in his 1857 *Les
Fleurs du Mal* (*Flowers of Evil*).
Much like Eugénie, Baudelaire
danced between worshipping
and reviling exotic beauty
and ostentatious wealth. One
of the best-selling editions of
this book of poetry captures
this feeling in its cover: a royal,
erotic, and piercing image of
a green and black orchid with a red throat. It is an invented flower, but
certainly one inspired by the cypripedium orchid alliance of paphiopedilums and phragmipediums—throughout time thought to be some of the
most anthropomorphized flowers.

In fact, Baudelaire invoked both Eugénie and the Second Empire
throughout *Fleurs du Mal.* Many poems were so racy, Baudelaire was
fined 300 francs for indecency. In spite of his critique of her reign, like
a few other authors, he appealed directly to the empress to reduce the
court's fine. Eugénie waited several months to reply—likely hoping to
skirt any negative press associated with Baudelaire—and reduced the fine
but did not absolve him of guilt. Although the poet had no direct contact
with the empress, his "Invitation to the Voyage" describes Eugénie's style
of interior décor and hints at dreams of congress with elite women in
their royal bedrooms:

The flowers most rare
Which scent the air
In the richly-ceiling'd gloom,
And the mirrors profound,
And the walls around
With Oriental splendour hung,
To the soul would speak
Of things she doth seek
In its gentle native tongue.

By late 1860, the empress needed time away from court. She was in deep mourning over the death of her sister from breast cancer and needed to get away from Napoléon III's now-constant sexual exploits. She decided to take a three-week independent grand tour of Scotland— a journey of connection with her matrilineal homeland. (For French royalty, "independent" meant traveling with merely two ladies-in-waiting, two gentlemen, and ten servants—all of the people one needed to take care of one's itinerary, hair, wardrobe, and various fashionable accoutrements.) She also visited her confidante Queen Victoria in an intimate stop at Windsor Castle. Other than her desire to escape into the arms of an understanding friend, by all accounts the centerpiece of the trip was Abbotsford in the Scottish borders, originally the country-side home of Sir Walter Scott (1771–1832).

Much like her childhood summer with Irving, the influence of Sir Water Scott (author of *The Lady of the Lake* and *Bride of Lammermoor*) on Eugénie's life had been multiform. Much like his books, his home was a pastiche of mythical references, architectural styles, and anti-quarian curiosities shot through with armor, exotic orchids, wild forests, and flanked by large bucolic kitchen and cutting gardens.

There Eugénie found Scott's library composed of several thousand volumes of history, law, literature, poetry, horticulture, and a signifi-cant section devoted to the dark arts. "The Wizard of the North" had been dead since 1832, but his place in Scottish culture and wider litera-

Sir Walter Scott's Abbotsford home (a castle, really) and kitchen garden
in Scotland.

ture was already enshrined. And quite the "picturesque pile" his home
was—Scott spent most of his life in enormous debt building the castle
of his dreams, referencing history and architecture when it suited him
and making up the rest along the way.

Eugénie also spent time in Scott's Chinese Drawing Room, a striking green chamber with crimson curtains, an imposing chandelier,
and a large hearth. Its hand-painted seafoam wallpaper, imported from

China in 1822, featured floor-to-ceiling pink peonies the size of dinner plates. Placed repeatedly throughout the design were also native Chinese orchids in everyday use, including white cymbidium species in a homey outdoor table arrangement and lining well-kept paths, as well as phaius species growing profusely around men playing the ancient game *Go*. Now this was a style to take inspiration from—Eugénie found the mix of cozy seating encircled by sumptuous exoticism and magical greenery intoxicating. She went on to similarly design her rooms in the Tuileries and at Fontainebleau on these themes.

Orchids reigned in Eugénie's greenhouses but were also a permanent feature of her interior décor. Paintings Eugénie commissioned reveal her varied use of orchids, often within a domestic setting. Her style, which became the style of the empire, was a mix of neoclassical clarity and rococo opulence. She leaned heavily upon Chinese, wider Asian, and sylvan aesthetics, with infusions of fairy magic. Childhood nature fantasy often came to life in her design choices.

Giuseppe Castiglione's 1861 painting of Eugénie's personal study reveals that like Sir Walter Scott's Chinese room, the space was covered in green silk with red velvet and gold accents. Masses of orchids take seat in an immense jardinière. A friend wrote that Eugénie's "writing-table was encircled by a crystal screen over which graceful climbing plants

Sir Walter Scott's Abbotsford Chinese Sitting Room wallpaper with cymbidium orchid (1822).

Castiglione, *Empress Eugénie in the Salon at the Tuileries Palace* (1868).

hung in festoons of verdure, giving her the appearance of being in some nook of tropical forest." In earlier years, the screen was made of gilt bamboo covered in ivy. The room was always full of flowers, and in large windows "winter gardens" were installed every year. She loved roses, carnations, and geraniums, but the light from the window was reserved for orchids, here likely oncidiums and dendrobiums. Upon her desk in the next room appears to be a vase of pink and white paphiopedilums.

The aesthetic, if a bit more formal, was carried throughout the Tuileries' semi-public spaces of the empress's *salons vert, rose,* and *bleu.* And in 1863, Eugénie went on to design her own Chinese room, or what she called her *Musée Chinois.* In its formation, we can see how far the British-French alliance had come in such a short time. Following failed trade negotiations with China, in 1860 the allied Franco-British armies organized a violent expedition into Beijing to reinstate European

ambassadors and open trade routes. The troops continued on past the
capital to the Imperial Palace, known as the Garden of Perfect Brightness or Old Summer Palace. It was not strategic militarily, but heavy
with cultural importance. There, British and French soldiers looted the
palace and gardens—then thought the pinnacle of Chinese architecture
and landscape design—and set fire to them. Empress Cixi, as we have
seen, was greatly affected by the gardens' destruction and its symbolism.

Three large lots of ransacked artwork, chinaware, ivory, and other
objects were addressed expressly to Empress Eugénie from one of her
generals; most of it became the base of her collection in her Chinese
museum at Fontainebleau. It remained a statement of France's imperial
power and continued commercial colonization of Asian nations after
a militarily enforced opening of trade with the region. The empress,
after all, had aestheticized colonialism, too.

Eugénie's story ends somewhere in between a happy fairytale and the
fate of Marie Antoinette—she was a fallen but continuously pampered
empress. In 1870, Napoléon III was defeated in the Franco-Prussian
War and forced to capitulate his seat as emperor just as revolution
broke out at home. He, Eugénie, and Prince Imperial Louis Napoléon
separately made their way to England.

Much of Eugénie's jewelry was later smuggled out of the Tuileries
and spirited to London in a diplomatic bag. The diamond *Coelogyne
cristata* brooch and other pieces were sold in 1873. The *London Daily
News* and *New York Times* reported about the lot that "for taste and
workmanship combined, with historical interest, it is not too much to
say that nothing like it has been offered for sale in England within living memory, if at all." They also exclaimed, "The triumphs and hom-
age these glittering gems have seen, and the atmosphere of splendor
in which they have been displayed, are known to all." (The coelogyne
brooch most recently sold in 2009 at Christie's of London for $375,000.)

The family had enough left from its wealth to live on comfortably—
comfortably enough to reside in a large estate in England and hire
several dozen servants. Napoléon III died in 1873, and their son died

abroad in 1879. From that moment, Eugénie, former fashion icon, wore black for the rest of her life. Victoria, too, had worn the dark colors of deep mourning since Albert's death in 1861.

During the last three decades of the century, Victoria and Eugénie's friendship grew—the two were a permanent feature of the English royal scene. The former empress regularly accompanied the queen to Windsor, Osborne, and her Scottish castles. Victoria affectionately asked Eugénie to stop calling her "Your Majesty" or "Madame"—"Why not 'sister' or 'friend'?—that would be so much more pleasant." One of Victoria's attendants wrote that in her advancing age, "The movement of the Queen, crippled though she was, was amazingly easy and dignified; but the empress, who was then sixty-seven, made such an exquisite sweep down to the floor and up again, all in one gesture, that I can only liken it to a flower bent and released in the wind."

COELOGYNE CRISTATA

Coelogyne cristata is a fragrant, medium-sized orchid, with cascading white blossoms and a golden streak on its lip. It can be a bit of a trick to provide it the winter cool-down it needs, but devotees of this orchid find it well worth it. In places that receive little frost, *Coel. cristata* can be mounted to encircle a tree trunk like a spectacular crinoline ball gown.

Orchid Details

— Place of origin: Himalayas in China, Nepal, and Bhutan
— Blooming season: winter and early spring
— Flowers last for: more than one month
— Plant size: 6–8 inches tall; leaves 12 inches long atop 4-inch pseudobulbs; can grow to the size of a small bush
— Flower size: 3 inches wide; inflorescences hold 5–8 flowers each
— Fragrance: banana, sweet fruit
— Plant habit: sympodial; leaves and elegantly drooping, flowers borne above the leaves

Orchid Needs

— Light: medium-high light with some sun; brightest in winter
— Temperature range: summer, 58°F–75°F; winter, 36°F–55°F
— Humidity: high year-round, 60%–85%
— Water: this orchid thrives in summer monsoon, but dry, cold winters (for at least two months)
— Growing medium: medium bark, perlite, and charcoal
— Fertilizing schedule: "Weakly, weekly" with a balanced orchid fertilizer in the growing season, then gradually reduced to none
— Seasonal changes: very heavy water through four months of summer; dwindling to nothing for a short winter
— Special requirements: To bloom at all, *cristata* must be cold, dry, and bright for at least two months. Near the end of this period, growers report that watering the orchid with handfuls of snow is a way to gradually increase its water. This is the one time when it's acceptable to "add ice" to your orchid!

Chapter 12

The Historian, the Actor, and the Healing of Orchids

O rchid growing, and especially hybridizing, is a practice that requires an outsized amount of patience. It can take years for a seedling to bloom—from cross-pollination to the flowering of a new hybrid, it is not unheard of to wait a decade. That decade is filled with, at a minimum, weekly maintenance of the growing plant. All of this is to say that it takes a person with longanimity to do more than dabble in orchids. You must put in the time, and still a great number of plants may die, and experiments may not bear the flowers one has hoped for.

One person with this kind of forbearance was the esteemed African American historian John Hope Franklin (1915–2009). He lived as a

John Hope Franklin holding a phalaenopsis named after him in the third-floor greenhouse of his home in Chicago, Illinois (1978). *STM-035851372, Kevin Horan/ Chicago Daily News. © Sun-Times Media, LLC, and Chicago History Museum.*

child under severe segregation and without running water or electricity in Oklahoma until he was a teenager, was nearly lynched at the age of nineteen, and yet became one of the greatest scholars of the twentieth century. Publishing eleven landmark books over his illustrious career, he reconceptualized the way Americans think about the history of slavery, the Civil War, and Reconstruction; helped write a legal brief for *Brown vs. The Board of Education* (the Supreme Court case outlawing segregation in public schools); and marched with Dr. Martin Luther King Jr. from Selma to Montgomery, Alabama, in 1965. And, yes, as he wrote in his autobiography, "living in a world restricted by laws defining race . . . challenged my capacities for survival."

Franklin was also a passionate devotee of orchids. Over the course of three decades, he cultivated nine hundred species from the six continents where orchids are native. Franklin, his wife Aurelia, and son John Whittington Franklin developed a fascination with orchids as a family in 1959, after a summer the historian spent teaching in Hawai'i. Franklin described the experience, "becoming acquainted with and attached to the cultivation of orchids . . . changed my life forever." On walks near the University of Hawai'i with Aurelia, the couple stopped to inspect every orchid they saw, whether growing in trees or blooming in neighbors' yards.

In addition to Hawai'i's beauty, they and their young son experienced fewer acts of overt racism there than they had in the US South and on the East Coast; they seriously considered moving to Oahu permanently. In the end, Franklin decided that Hawai'i at that time was "too far from the center of the fight" for racial justice and the family returned to New York. They brought terete vanda orchids home with them and proceeded to build a three-shelf window grow space in their house in Brooklyn. While John taught at Brooklyn College, Aurelia joined the Faculty Wives' Club and befriended Pat Withner, wife of botanist and "orchid legend" Carl Withner. Through the Withners, the Franklins had ready access to orchids and cutting-edge orchid research to feed their hobby.

Franklin's love of orchids grew, and the orchids followed the family as John took a post at the University of Chicago in 1964. The next year, the family built a 9 foot × 11 foot greenhouse on the flat roof of their home. They grew a bit of everything—odontoglossums, phalaenopsis, ascocentrums, equitant oncidiums, cattleyas, and sophronitis. Soon, Franklin acquired an orchid import license so that he could bring orchids into the country as he returned from numerous international speaking engagements. He acquired vanilla in Zanzibar, laelias in Costa Rica, *Broughtonia sanguinea* from Jamaica, and many native species in Brazil and Cape Town.

By 1980, the Franklins' orchid collection had grown substantially,

Ansellia africana. Warner, Williams, and
Moore, *Orchid Album* (1889).

and they moved to Durham, North Carolina, for John's new position at the National Humanities Center and later at Duke University. There, they built their "dream greenhouse," a 17 foot × 24 foot structure with graduated stair benches around the perimeter and down the center. With begonias and ferns flourishing in the humidity underneath the benches, on top they grew *Psychopsis papilio*, *Dendrobium nobile*, *Cattleya* Molly Tyler, *Cattlianthe* Orchidglade, "one of the great orchid hybrids of the world," John explained, as well as the striped yellow and brown *Ansellia africana* from the Nile Valley. A variety of dendrobiums, vandas, phalaenopsis, and cattleyas rounded out the large collection, and John and Aurelia were active members of Durham's Triangle Orchid Society for several years.

In the United States' bicentennial year, 1976, a phalaenopsis was named after John Hope Franklin. The complex hybrid moth orchid, with white petals and a red lip, was named by Hermann Pigors, owner of Oak Hill Gardens at the time, in Dundee, a township outside of Chicago. Pigors, who attained professional orchid training in Germany but has lived most of his life in Illinois, remembers Franklin as a thoughtful friend (and a good customer). The orchid grower went on in 2003 to create and name a large lavender cattleya with ruffled petals after Franklin. Aurelia had an orchid named for her as well; *Phalaenopsis* Aurelia Franklin was created by Carter & Holmes Orchids in South Carolina in 1998. It is a small, freckled, bright yellow orchid—John attested that it was his favorite orchid, and once joked to a reporter that like its namesake the phalaenopsis was "long-suffering and tolerant." His other favorite, the cattleya Pigors named for him, was also like its moniker in that it was "big and ungainly." Franklin often made connections between orchids and his profession as well. He found both orchids and American history were "full of challenges, mystery" and easy to fall in love with. "To grow orchids, you have to be persistent, patient," he warned. "And to do the right kind of history, the kind of history I think is worth doing, of course, you have to be persistent *and* patient and work hard."

Franklin continued traveling the world serving on national delega-
tions and giving lectures on American history, returning with orchids
in his luggage. On one such trip, a graduate student he had employed to
care for his collection back home accidentally knocked over one tiny,
rare, red-petaled orchid and broke its pot. When told of the accident,
the historian consoled the distraught student, "Don't worry. Orchids
are hardy plants," the gesture underlining Franklin's warmth and keen
ability to sense the connections between the best qualities displayed by
humanity and the plant kingdom. Professor and author Cathy Davidson, a colleague of Franklin's at Duke, remembers their mutual love of
orchid growing as similar to their dedication to civil rights: "Orchid
growing is now my John Hope Change Metaphor: if the conditions are
absolutely right, change is easy. Under ordinary circumstances, it takes
enormous, dedicated, constant, and attentive labor to get something
unique and beautiful to bloom."

John Hope Franklin with blooming cattleya in Durham greenhouse.

The year that he had a phalaenopsis named after him, in a review he wrote about a new orchid book, Franklin described the great pleasure he took in his orchid hobby. He detailed his love of "the history, the provenience, the uses, the drama, color, form, and even the sex life of this hauntingly beautiful family that numbers more than 30,000 species growing wild all over the world." Orchids deeply rejuvenated a man otherwise preoccupied by terrible histories and intractable politics. Franklin approached orchid growing like he did everything else in his life: with thoughtfulness and hard work. And as President Obama eulogized Franklin in 2009, "because of the life he lived, we all have a richer understanding of who we are as Americans and our journey as a people."

⁓

Worlds apart but nonetheless contemporaries, John Hope Franklin and the actor Raymond Burr were professional men who had worked very hard to achieve their place in an America that promised self-fulfillment and respect. Yet the 1950s and later decades—however golden they may have seemed on screen—were nevertheless a dangerous time for black people and gay people in the United States. Franklin and Burr waded through overt and covert racism and homophobia, respectively, as part of their daily lives and careers.

Burr's first roles in Hollywood were as the "heavy" in noir and cowboy films. His height and girth fit these parts in the 1940s, as well as his resonant voice. Yearning for leading man roles but more often cast as a character actor or villain, he enthusiastically followed the film industry practice of inventing maudlin details of his life for reporters, and then embroidering them for years to come. First, he had one dead wife, then two; later a deceased baby was added to the yarn, and then a third dead wife. He claimed to have traveled the world during years in which there is much evidence that he was in the United States. As part of a

Hans Erni, *Raymond Burr with Orchids* (c. 1985).

cover to conceal his sexual orientation, he felt obligated to repeat many of the false details of his backstory until his death.

Burr's most memorable roles were the chilling Lars Thorwald in Alfred Hitchcock's 1954 classic *Rear Window*, and the titular star, an unflappable defense attorney, in TV's *Perry Mason* (1957–1966). Showing off his masculinity and conveying heterosexuality in order to continue his career, he became a regular on USO tours and was set up on public dates with beautiful women, including the much younger Natalie Wood. Much like Montgomery Clift and Rock Hudson, Burr's employment depended on the machismo he projected. And so, in the most queer-phobic of times, Burr was always under the public microscope, although his homosexuality was something of an "open secret" with his colleagues. Luckily, he was mostly left alone by the gossip magazines; the FBI, under the direction of J. Edgar Hoover, however, kept a file on "sex deviate" Burr.

For both Burr and Franklin, their love of orchids provided a regular respite before stepping back into the fray. Psychiatrist Sue Stewart-Smith in *The Well Gardened Mind* reasons that working with plants becomes a kind of daily therapy for people under long-term stress. "A garden gives you a protected physical space, which helps increase your sense of mental space." Gardens can also be places of solitude, where one can think quietly, even forming a sort of meditative space. "The more you immerse yourself in working with your hands, the more free you are internally to sort things out and work through them." And so, throughout time, indoor and outdoor gardens have served as refuge for people and groups coping with various types of prejudice and trauma.

Franklin and Burr also came of age in an era of terrific advancements in orchid cultivation. In 1922, plant physiologist Lewis Knudson at Cornell University published the results of experiments sowing cattleya seed on agar in sterilized flasks. Previously, growers working with orchids relied upon sprinkling their dust-like seed at the base of mother plants so that it could benefit from mycorrhizal fungi in the media during germination. Seed mortality was very high, and thus orchids remained exorbitantly expensive. With Knudson's method, growers were able to germinate and raise thousands of orchids from a single seed pod. This benefited the trade in orchid species—growers did not have to rely upon plant division to increase plant numbers—but it utterly revolutionized orchid hybridization. Crossing two orchids results in an orchid with traits from both parents, and like human children, each seed in an orchid pod varies in its genetic presentation. Paphiopedilum breeder Ross Hella explains the results more colorfully: "hybridizing orchids is like wood carving with a shotgun." If you continue crossing hybrids, complex new orchids are born. With reliable germination and speedy growth using this new method, orchid fanciers had a lot more flower stock to play with and collect.

A second breakthrough in orchid cultivation occurred in 1964, when a French company, Vacherot and Lecoufle, cut a tiny tip of a new growth on an orchid and placed pieces of it on a rotating wheel in sterile nutrients. Within a few months, a few hundred orchid protocorms grew and developed into plants. This type of tissue culture, also called meristem culture, allowed orchid growers to effectively clone plants. It meant that exact matches of highly sought-after species and hybrids could be sold for lower prices and reach more gardeners. Instead of a fine variety of cattleya selling for hundreds of dollars, by the late 1970s, a grower could acquire a meristem-grown orchid for $10–$15.

Post-war technology and global transportation networks expanded orchid access, too. In peacetime, the leading aircraft manufacturer Boeing began selling its cargo jets and transport planes (the 747 and 707) to airline companies. Orchids could now be shipped around the world in a matter of hours—grown and even brought into bloom in a tropical location but sold thousands of miles away. Seed-grown orchid

Orchids were fashionable in post-war America. *American Orchid Society Bulletin* (1946).

technology and global transportation networks especially revolutionized the phalaenopsis trade, as a phalaenopsis can be grown from seed
to flowering size in three years (by comparison, cattleyas take about
seven years to bloom; cymbidiums even longer). Orchids continued in
their role as bold fashion statements and eye-catching centerpieces at
fancy events though mid-century.

Franklin and Burr were both nature lovers from an early age, and
although their lives differed greatly, both took solace in orchids. Raymond Burr was a shy child, and without many friends, he spent a lot of
time in the garden with his mother. Ostracized, bullied, and taunted
with chants of "fatso," he would often escape his San Francisco Bay
area school and sneak quietly into neighboring gardens, safe and alone
with the flowers. By the time Burr was a young adult, the United States
was deep in the throes of the Great Depression. Hoping to work with
plants—to land a paying job to support himself and his mother—he
joined the Civilian Conservation Corps. Stationed in California, his
tasks didn't involve plants so much as trenches, fences, and concrete.
But he enjoyed it and began to think about an alternative and perhaps
simultaneous career—acting.

Although Burr always came across as manly, his fondness for orchids
may have raised eyebrows—effeminacy through flowers was thought
to be the central hallmark of a gay man's character throughout the
twentieth century. For decades, Oscar Wilde, the Irish poet and playwright, who was found guilty of and imprisoned for having a homosex-
ual affair; Robert Hichens, the popular British writer of the first half of
the twentieth century; African American poet Langston Hughes; and
several other authors published texts in which decadent floriculture
connoted queerness. As early as 1904, Howard O. Sturgis, an American
living and writing in Britain, described an attractive gay man as "some
strongly scented hothouse flower, white with a whiteness in which there

was no purity, and sweet with a strong sweetness that already suggested some subtle hint of decay." While fiction often depicted queer people as threatening, some authors also worked to normalize queerness. Marcel Proust, the gay author of *Remembrance of Things Past*, explicitly tied gay men to orchids in his 1921 *Sodom et Gomorrhe*, describing a male admirer gazing at a gay man who "struck poses with the coquetry that the orchid might have adopted on the providential arrival of the bee"— that is, in a highly stylized and yet natural way. By the 1920s and '30s, gay men were pejoratively labeled a variety of flowers, including pansy, daisy, and buttercup, as well as "fairies." Nature and garden references were so extensive that stereotypes then became generalized—gay men slandered with, and then repossessing with pride, the catch-all phrases "horticultural lads" and "horticultural gents."

Orchids may have caught on with queer people at this point in history because the flowers so often confound our expectations—orchids are often described as bold rather than delicate, and the thick, waxy blooms of cattleyas, especially, have been understood as hard and active, rather than soft and passive, flowers. Gay men might have taken to orchids because of their generally erotic flower structures, or perhaps because *orchis* is Greek for testicle. Orchids in general, along with carnivorous plants and a few other shocking tropical flowers, were often described as "queer" in fiction and nonfiction by the late nineteenth century. Notably, this is also when "queer" became a slur to describe anyone not perceived as heterosexual.

A closeted Burr played his Hollywood roles well, proved a successful actor, and was a very wealthy man by the late 1950s. He bought a house in Malibu and started speedily acquiring orchids and making his own hybrids. Because he wanted warm-growing orchids from other countries that were unavailable in the United States, he soon needed a license for importation. In part to secure the license, he founded Sea God Nurseries and ran it with his romantic partner, Robert Benevides, whom he met in 1960.

The couple eventually added several more growing locations, and moved from Malibu to the Hollywood Hills, where they built thirteen greenhouses and filled them with thousands of their own hybrids. By the late 1970s, Sea God was turning a profit, and during a lull in his acting career, Burr decided to focus on orchids and flowers more generally. He made appearances to promote his orchids, and invested early on in Teleflora, still today one of the largest flower delivery services in the world. In an interview, Burr revealed "the fact is, flowers are my passion. I believe that by enabling people to send flowers . . . we're helping the country, I hope aesthetically and culturally."

Burr and Benevides eventually had orchid nurseries in California, Fiji, Hawai'i, and the Azores. Burr considered Fiji his second home and developed not only a nursery but also an orchid garden for visitors, calling it the Garden of the Sleeping Giant. It remains Fiji's largest orchid collection, focused on cattleya hybrids and the orchids of Asia.

The couple, over the course of their orchid growing career, created over two thousand hybrids, registering almost three hundred of them. They had a decided emphasis on cattleya alliance breeding but experimented with vandas and phalaenopsis as well. Burr was especially inspired by orchid hybridizers' perennial quest for creating "blue" blooms. True blue is rarely found in flowers; the closest most orchid growers come to blue is the tint of Dauphin's violet—a purple the eye can perceive as light blue in some light and in comparison to darker purple flowers. In orchids, this trait is technically labeled *coerulea* (Latin for blue) and, by 1985, Burr achieved success with it in his hybrids, when his *Cattlianthe* Sea King was noted as one of the best *coerulea* cattleya cultivars by the American Orchid Society.

Burr adopted a few themes in his orchid naming; emphasizing the Sea God brand, he and Benevides created *Rhyncholaeliocattleya* Pieces of Eight, *Phalaenopsis* Doubloon, *Cattleya* Nymph, *Cattlianthe* Harvest Tide, *Rlc.* Sea God Gold, and *Ctt.* Sea King. Many orchids were named for jazz singers (*C.* Nat King Cole, *Ctt.* Ella Fitzgerald, *Rlc.* Lady Day), islands in the Azores, and Los Angeles landmarks. They also named orchids after Fiji's first lady, Adi Lady Lala, as well as Queen

VANDA COERULEA LORD ROTHSCHILD'S VARIETY

Burr and Benevides experimented with
blue tones in orchids. *Vanda coerulea*.
Warner, Williams, and Moore, *Orchid
Album* (1897).

Salote of Tonga, and Princess Grace Kelly of Monaco. The theme he was best known for, however, was giving hybrid orchids the names of his film and television co-stars, including cattleyas named for Alice Ghostly, Betsy Jones-Moreland, Charles Macaulay, Jean Simmons, Laura Thayer, Lenore Shanewise, and cattleya alliance hybrids named for Barbara Anderson, Barbara Hale, Florence Henderson, Lady Barbara Stuart, and Molly Picon.

Sea God Nurseries had developed into a serious mail-order business by the early 1980s, valued at $2 to $3 million dollars. But with the end of acting in TV series, Burr wanted to get out of Southern California. He and Benevides planned a move from Hollywood to a large parcel of land they owned in Healdsburg in northern California's Sonoma County. There, they would expand their horticultural and farming interests, take up grape cultivation, and found a successful winery.

Upon the move, Burr donated a large portion of his orchid collection—some 3,200 cattleya alliance hybrids—to California Polytechnic University, Pomona, in 1981. He also donated a 25 foot × 50 foot H-frame greenhouse and equipment for cultivating and breeding orchids. Burr additionally gave Cal Poly 3,600 cattleyas, laeliocattleyas, schombocattleyas, and other similar orchids in 1983, his donations to the college totaling almost $150,000 (about $440,000 today). Cal Poly's development director wrote to Burr, "generations of students will profit through your generosity, and we want you to be assured that your gift is greatly treasured."

But the actor still had enough orchids to fill ten truckloads in the move to Healdsburg. In 1986, *Orchids* magazine listed Sea God Nurseries among the most significant cattleya alliance hybridizers in the world. By that time, Burr was also well known for developing several interesting primary hybrids with *Encyclia cordigera*, an easy-growing chocolate-scented orchid.

Jackie Lacey, former president of the American Institute of Floral Designers, explains the long-standing history of flower appreciation

Raymond Burr in one of his greenhouses at his home in Hollywood, California (1977). *Ron Galella/Ron Galella Collection via Getty Images.*

for queer communities: "Flowers have been a part of the LBGTQ+ community for centuries . . . flowers symbolize what we've been through together and look forward to together." Among other flowers, orchids have been a part of healing the past and present for aggrieved communities in the United States.

More generally, orchids fulfill many human needs, especially when we are in pain or mental anguish. The desire to create something beautiful, to nurture living things, to busy one's hands and free one's mind for a moment, to get lost in the intricacies of nature . . . all these are achieved through a little quiet time in the presence of plants. Sue Stewart-Smith describes that green spaces of all sizes do even more than that: "Like a suspension in time, the protected space of a garden allows our inner world and the outer world to coexist free from the pressures of everyday life. Gardens in this sense offer us an in-between space which can be a meeting place for our innermost, dream-infused selves and the real physical world."

PHALAENOPSIS JOHN HOPE FRANKLIN

Phalaenopsis John Hope Franklin, like American history, is a complex hybrid— it is a flower evolved from several generations of resilient stock. Although this specific hybrid is difficult to obtain today, semi-alba complex phalaenopsis with large white petals and a red-pink lip like the historian's own orchid  are easily found and are one of the very easiest orchids to grow. Simmie Knox painted *Dr. John Hope Franklin with Orchid* in 1996.

Orchid Details

~ Place of origin: this hybrid's grandparent species are endemic to Indonesia, the Philippines, Taiwan, and the surrounding area in the South China Sea

~ Blooming season: in the northern hemisphere, peak blooming season for complex hybrid phalaenopsis is winter, but they can be found in stores year-round

~ Flowers last for: six months if kept shady and cool

~ Plant size: its tall, slender inflorescence often makes the plant more than 2 feet high

- Flower size: 3 to 4 inches across
- Fragrance: complex hybrid phalaenopsis have been bred for color and size, not for fragrance, so very few carry any scent
- Plant habit: monopodial—medium-sized leaves grow in a ladder formation; its lofty, graceful inflorescence can be trained when growing

Orchid Needs

- Light: low light, shady conditions
- Humidity: 60%-70% humidity ideal, will cope with less
- Water: water regularly, but do not let growing medium become waterlogged; healthy roots turn green when wet, silver when dry; allow roots to dry but not become desiccated
- Growing medium: sphagnum moss and medium or large fir bark
- Fertilizing schedule: "weekly, weakly" with one-quarter dose of balanced orchid fertilizer
- Seasonal changes: nights in the mid-60s°F, days in the mid-80s°F; without regularly experiencing a 15°F day/night difference, the orchid may need a week with nights in the 50s to initiate a spike
- Special requirements: none

GUARIANTHE BOWRINGIANA VAR. COERULEA

Unfortunately, most of Raymond Burr's hybrids are unavailable today, but the quest for excellent blue cattleya alliance hybrids continues. *Guarianthe bowringiana var. coerulea* was widely used by Burr, and is a reliable parent species breeders use today, offering vigorous growth and large clusters of sparkling flowers on each pseudobulb.

Orchid Details

- Place of origin: Guatemala and Belize
- Blooming season: in the northern hemisphere, it is an autumn bloomer, peaking in October and November
- Flowers last for: 2 to 4 weeks

- Plant size: size is determined by quality of care. Exceptionally good care can produce a very tall plant, its pseudobulbs alone 20–30 inches; mature flowering plants can range from 2 to 4 feet tall and just as wide
- Flower size: can have fifteen flowers or more per spike, each bloom 2–3 inches wide
- Fragrance: varies by the cultivar; some growers report no fragrance, others report a heavenly floral scent
- Plant habit: sympodial; tall and erect; blooms form on straight, upright spikes

Orchid Needs

- Light: bright light, can take some sun
- Humidity: 80% all year ideal, will tolerate less
- Water: 2–3 inches per month June through October; dwindling to very little water February through April
- Growing medium: a chunky, airy mix of large bark and perlite
- Fertilizing schedule: use balanced fertilizer with micronutrients at half strength weekly, less or none in the dry season
- Seasonal changes: winter nights in the high 50s°F, winter days in the high 70s°F; summer nights in the high 60s°F, summer days in the high 80s°F
- Special requirements: winter is the brightest season for this orchid, but it is also cooler and quite dry in its native habitat

Orchid Art and Conservation

The scale of the Amazon River and its wider basin in South America is almost too large to comprehend. The sweeping watershed covers more than 40% of the continent; the Amazon and its hundreds of tributaries—if uncoiled and linked end to end—would twice encircle Earth. Its riverine ecosystems, wending their way through Ecuador, Colombia, Peru, Venezuela, Bolivia, and ultimately to the Amazon's mouth in northeastern Brazil, host more than one million indigenous people and three million species of animal and plant life. Several thousand orchid species live here in rain forests, floodplains, mountains, and savannas, adding to one of the most biodiverse regions on Earth. Amazonia is the single largest tropical rain forest we have left.

The orchid industry has, unfortunately, contributed to the long, dark history of rain forest destruction worldwide. Driven by collectors' orchidelirium in the nineteenth century, plant hunters collected specimens throughout the Amazon and elsewhere, sometimes destroying large sections of forest in the process. Just one of these plunderers, Englishman Albert Millican, traveled to South America several times in the 1880s and 1890s, collecting any orchid he thought would sell. Once in Colombia seeking oncidiums growing high in the canopy, he instructed hired men to chop down any tree that hosted orchids. In two months, they "had secured about ten thousand plants, cutting down to obtain these from some four thousand trees, moving our camp as the plants became exhausted in the vicinity." History holds additional reports of orchid hunters—attempting to corner the market on certain rare orchid species—who set fire to areas they had just ransacked.

But before these abhorrent facts turn off would-be orchid lovers, we should also acknowledge that for generations and yet today orchid enthusiasts have proven crucial in the restoration and preservation of the Amazon and other orchid-rich regions of the world. Painters, naturalists, orchid societies and charities, professional growers, and home growers are now vibrant players in efforts to conserve the diverse ecosystems in which their favorite orchids thrive.

One Victorian artist, Martin Johnson Heade, exhibited an incipient conservationist ethic in several dozen orchid and hummingbird paintings he executed in the 1870s and 1880s. Best known as an American landscape painter, he was influenced by Darwin's work and argued for the legal status of species after reading *Origin of Species*. Heade's vocal support of wildlife and landscape conservation would not fully emerge until the 1890s, but his advocacy likely germinated during several trips to Central and South America in the 1860s and 1870s. Visiting Brazil, Nicaragua, Colombia, Panama, and Jamaica, he was sure to have seen orchids in his journeys but spent the most time with the flowers in the hothouses of the Royal Botanic Gardens, Kew, and at friends' estates in Britain and the United States. With these experiences, Heade began to set large cattleyas and delicate hummingbirds on his canvas within detailed and mercurial landscapes. Mixing elements of still life with landscape art, his work invited viewers to become companions to nature. Unusual for the age, his works portrayed jungle scenes with curiosity rather than a view toward the exploitation of their resources. Heade very well may have been winking at Darwin in these works as well. Visual attractiveness, he had learned from the famous naturalist, was a major component of natural selection in the lives of both the jewel-toned hummingbirds and the glittering orchids he painted.

A century later, the desperate need for evocative visual representation within a crisis of comprehending the destruction of the Amazon was made clear by Prince Philip, Duke of Edinburgh. "The bare statistics about the destruction of the tropical rain forests are simple statements of fact," he wrote, "They cannot hope to convey the full

Martin Johnson Heade, *Orchid with Spray Orchid and Hummingbirds* (c. 1875–1890). *Photograph © 2022 Museum of Fine Arts, Boston.*

Margaret Mee sketching in the Amazon rain forest, Rio Negro, Brazil (1988).

scope of what is rapidly becoming a major global tragedy." Recounting then-recent worldwide devastating hurricanes, floods, and famines, he noted in terms of damage to the natural environment, their combined effects did not compare with the annual destruction of the Amazon rain forest. "In 1987," he wrote, "nearly 80,000 square miles . . . went up in smoke." The Duke of Edinburgh then described the lasting impact of artists like Margaret Mee, a woman who had spent the better part of her life using her talents to paint species of the Amazon and disseminate knowledge of them in Brazil, as well as Britain, wider Europe, and America. She combined these compositions with careful observation and daily records of small- and large-scale change in the Amazon over a thirty-four-year career on that river. Her work presented a slowly unfolding catastrophe in a way humankind could process—and perhaps more importantly—feel. Prince Philip affirmed that the Amazon needed the "eyes and the talents of an artist to make that scale of destruction comprehensible to the human mind." The lush natural environments in Mee's art underlined her protest of industrial clear-cutting and catapulted the preservation of the Amazon to an issue of global importance.

Born in England in 1909 to a nature-loving family, Mee attended art schools in London and graduated with a degree in painting in 1950. She and her spouse, fellow artist Grenville Mee, were lovers of travel. The pair moved to São Paulo, Brazil, in 1952 to care for Margaret's ill sister, intending to stay only a few years. Grenville was soon established as a busy commercial artist, and Margaret taught art classes at a British school in São Paulo. The couple enjoyed the city's parks and took hikes in the surrounding hills with their sketchbooks. On one trip, spotting a castor-oil plant in full bloom and entranced by its odd textures and shapes, Margaret decided on the spot to "put aside all other ideas and began sketching and painting flowers" with every spare moment. Margaret soon became so enamored of Brazilian flora in the city that she was driven to seek out more blooms in the vast surrounding jungle.

By 1956, Margaret and a Dutch friend and fellow teacher mounted an expedition to the River Gurupi. Flying north on an aging World War II aircraft from São Paulo to Belém (1,800 miles, or 2,900 km), they then caught a slow wood-burning train and a twenty-passenger river boat to their true embarkation point, the headwaters of the Gurupi. Hiring a local guide and small boat, the women soon found there were "flowers to paint in plenty." They encountered huge white sobralia orchids, *Rodriguezia lanceolata* throwing long sprays of cupped pink flowers, and the deliciously perfumed *Prosthechea (Encyclia) fragrans*, a cockleshell orchid with white petals and a magenta-striped lip. Throughout the journey, they slept in hammocks, walked miles in the jungles

Galeandra devoniana.
Paxton's Magazine of
Botany (1841).

looking for flowers to sketch, nearly starved, and encountered the Tembé and Urubú people.

Margaret returned from the trip having perfected her flower painting. Mee most often sketched, and sometimes painted orchids in situ, attempting to capture the plant in life size and as accurately as possible. When she could not finish painting, she kept detailed notes as to the color of the flowers and leaves. She took some of her work—sketches of the few plants she couldn't identify—to experts at the São Paulo Botanic Institute. Because of her attention to detail and precise color matching, they were able to identify them and encouraged her on in the work. The institute also arranged a small exhibition of her work, which then led, in 1958, to her first important exhibition in Rio de Janeiro, where seventy-six paintings from the trip were on display.

At the Rio show, she met and befriended famed Brazilian landscape architect and painter Roberto Burle Marx, a plant lover himself. Mee on later trips continued collecting species to take home. She gave the living orchids, bromeliads, and other plants to Marx and the Botanic Institute for their plant collections. Over a long friendship, Mee and Marx used their growing stature in the art world to advocate for conservation of the Amazon basin.

On her second trip to the jungle in 1962, this time invited to the

Gongora quinquenervis. Curtis's Botanical Magazine (1833).

Mato Grosso (translated as "denser scrubland") by Brazilian ethnographers Harald and Vilma Schultz, Mee encountered numerous sweetly perfumed orange and pink galeandra orchids, lime green *Catasetum fimbriatum*, and the spotted yellow *Cycnoches haagii*. In this and many other trips, some of the plants she deposited with the Botanical Institute were species unknown to science. Scientists and the wider world

Cattleya violacea
Rio Curni, Amazonas

began to come to the realization that unknown species were disappearing from the Amazon faster than they could be cataloged.

Mee's next trip centered on a location suggested by her friends at the institute—the Amazon's largest tributary, the Rio Negro. There she sketched *Catasetum barbatum*, a green orchid with a large furry white beard, and *Cstm. appendiculatum*, a spotted yellow orchid with red whiskers. At fifty-five years old, Mee may not have been comfortable with howler monkeys roaring in the night, thousands of ants crawling in her mouth and eyes, and eerily screeching birdsong, but she was in her element. At Christmastime, she was invited by six Tucano people and several local nuns to join them for a day hike to the Serra de Icana, a "shadowy and dark" forest where she was ecstatic to come upon "the outstanding find, the rare *Gongora quinquenervis*—such spectacular purplish mottled flowers. Wonderful!" Later, along the River Uaupes, she found *Epidendrum ibaguense* with tall, showy multifloral spikes with tangerine-red flowers. The plants she brought home from this trip were again divided between the institute and Roberto Marx; new species of bromeliad and philodendron were identified.

With travel support and a photographer provided by the National Geographic Society, in 1967 Mee journeyed to the Cerro de Neblina—the Mountain of Mist—in far northern Brazil on the border with Venezuela. Headed up the Rio Negro, Mee found that as they got further upstream, the rivers became more fascinating. Jara palms grew along the banks, "their fibrous stems making wonderful homes for dozens of epiphytes." She found profuse groups of fragrant orchids. The delicate white blooms of *Brassavola martiana*, maroon and white candy-striped *Galeandra devoniana*, and cerise flowers of *Cattleya violacea* "were so brilliant they gleamed in the trees." Mee was delighted to encounter *Cattleya violacea* on later journeys as well, and in 1981 painted one of

(facing page) Margaret Mee, *Cattleya violacea* (1981). © *Estate of Margaret Mee and the Board of Trustees of the Royal Botanic Gardens, Kew.*

Margaret Mee,
Sobralia margaritae
(1977). © *Estate of*
Margaret Mee and
the Board of Trustees
of the Royal Botanic
Gardens, Kew.

her most striking works on the subject. She captured the orchid in its natural habitat—a flooded igapó forest—with aquatic aroids, heliconia, and a red clusia completing the scene.

Returning from her trip with National Geographic, she brought one new plant species to light, and sadly, came down with infectious hepatitis. Almost three years later, Margaret was still in recovery, and she and Grenville had moved to Rio de Janeiro. In this time away from the Amazon, she was still able to paint and work with an old friend, Brazilian botanist Guido Pabst, illustrating and assisting in the assembly of his two-volume monograph on Brazilian orchids—by far the most comprehensive text on the subject. By this point in Brazilian history, new highways were in constant development through the Amazon, and Mee and her circle of friends were deeply worried about the fate of the forest. Mee and Marx continued to speak out about landscape conversion, but the Brazilian government controlled the national press, so foreign outlets were more sympathetic to their cause.

At the end of a trip in 1971, Mee was in Maués in the northern state of Amazonas only to experience one of the most difficult moments of her life on the Amazon River. "The forest was destroyed for miles around," she mourned. "Burned giant trees, stripped of vegetation and epiphytes, stood gaunt and white-scarred-black on arid soil . . . As I looked on this desert I tried to block from my mind a vision of the Amazon of the future." Fortunately, now the 2,500 square mile area around Alto Maués Ecological Station is a nature reserve; many other areas Mee visited were not so lucky.

Mee was only busier and more dedicated to the cause as the years went on. She attended the 1972 World Orchid Conference in Medellin, Colombia, and felt her contributions in the sessions on tropical America were successful. And on a trip to the Rio Urupadi, she came across an orchid she had never seen before. The water was high that day, and Mee and her guide "were trapped many times in pathless forests" on the small boat. Eventually they made it to the Rio Amena,

> and thence into a lovely *igarapé* [area navigable by canoes] where low palms growing in the water covered it with a blue haze and from their midst rose trees like columns in a Gothic cathedral. I scanned these trees for epiphytes, and on one of the tallest saw, with excitement, a huge wreath of orchids encircling the trunk.

The plant was not blooming at the time, but Mee knew it was a sobralia, and thrilled that it was surely "a giant in that genus." She was able to take a piece of the unknown orchid back with her. She delivered it to Guido Pabst, who couldn't identify it either. Soon the stately orchid bloomed with cream and yellow-green petals and a striped lip. Pabst published *Sobralia margaritae*'s first scientific description in 1977, naming it after Margaret, of course.

Coming home from a trip in 1982, Mee summed up what she often felt at the end of her journeys. "Tomorrow is a sad day of departure from the glorious River Negro and I shall be back in the world of turmoil, pollution and politics. When shall I be able to return?" she wrote wistfully. "At the moment the sun is sinking, sending a pink glow over the

Madeline von Foerster, *Orchid Cabinet* (2014).

sky and reflected gold in black water. Rapidly it is changing to a glowing brilliance against the shapes of birds of darkness." Two years later, at the age of seventy-four, she protested bauxite mining and pollution dumped into the Lago Batata. She also continued to mount regular international exhibitions and grant interviews to reporters from media outlets around the world about the precarious state of the Amazon rain forest.

Until her death by car accident in 1988 while in London, Mee had staged fifteen journeys through tributaries of the Brazilian Amazon, usually accompanied by a single river guide. Tireless in the world of art and botanical illustration, she was also an indomitable advocate for its conservation. Harvard professor Richard Schultes wrote that her paintings had given "powerful impetus to the growing outcry against the uncontrolled devastation" of the rain forest. In several respects, "Margaret, the quiet and unostentatious voice of the wilderness can be credited with one of the loudest voices for conservation."

With ongoing destruction of orchids' natural habitats everywhere, some of the most evocative modern painters like Madeline von Foerster (1973–) not only push the limits of symbolic representation in their paintings but also accentuate the human connection to the natural world. Working within the theme of conservation, von Foerster's oil painting *Orchid Cabinet* (2014) depicts eight endangered orchid species clinging to a carved ebony Mother Nature encased in a crate bound for the United States. *Phalaenopsis stobartiana*, a critically endangered species in China, blooms to the right of her face, while *Houlletia tigrina*, endangered in Central and South America, springs from her hair. A blue *Vanda coerulea* from Indo-China nests in a severed hand. An endangered Brazilian *Cattleya labiata* emerges from a drawer in Mother Nature's cabinet—invoking Western collectors' *wunderkammern* ("cabinets of curiosity") so popular from the Renaissance through the nineteenth century. Here, Africa's *Ansellia africana*, the Philippines' *Paphiopedilum*

fowliei, Taiwan's *Cypripedium formosanum*, and Madagascar's *Angrae-cum sesquipedale*—all vulnerable to critically endangered—also emerge from the cabinet, along with beetles, Darwin's moth, and a drawing of mycorrhizal fungi so critical to orchids' survival in the wild.

Born in San Francisco, von Foerster studied art in California, Germany, and Austria. Now living in Germany, the artist described her passion to an interviewer in 2009: "The human/nature relationship is the leitmotif of my art. I frankly feel this is the most important topic of our time, and our failure or success in regulating our exploitation of nature will be how future generations define us." Much of her work carries a trenchant environmental message—in *Orchid Cabinet*, the ravages of a global history of plant smuggling are on display, Mother Nature a parcel to be shipped and sold, the figure's direct gaze at the viewer a demand to acknowledge trafficking in endangered species. Yet, as unsettling as von Foerster's work can be, it is also undeniably attractive. This is in fact the artist's goal, stating that while her work confronts the most difficult aspects of ecosystem decline and global climate change, she tries "nevertheless to create beautiful paintings, which provoke reverence and contemplation." In this way, her art is intended to help foster the shift in consciousness which is needed to "not only to save the planet, but ourselves."

Utilizing the Mische technique, one developed by the Flemish Masters five centuries ago, von Foerster uses alternating layers of egg tempera and oil glazes. She contends that it works best for artists who carefully draw and plan their pieces. The technique's arduous reputation is representative of the hard work required of converting society to live more sustainably—von Foerster attests that "we are most alive when we are pushing for something beyond our capabilities and comfort zone." While we may often feel that all magnificent wild places on Earth are corrupted, denuded, and dead, it is a deliberate daily choice to acknowledge deforestation and climate change and still attempt to improve the chances for biodiversity on the planet. "Even as there is always infinite cause for despair and regret," she confided to the same journalist, "there is infinite possibility for grace, even bliss." And for all the research and thoughtfulness poured into her work, Madeline doesn't just paint; for years she's donated a percentage of her print sales to Tree Sisters, Save Vietnam's Wildlife, WildAid, and Sea Shepherd Conservation Society.

The orchid industry's past is riddled with the decimation of orchid species as well as the destruction of their native ecosystems. Yet today, it is orchid growers, orchid societies, scientists, artists, and directors of botanical gardens who are at the frontlines of orchid conservation efforts worldwide. Seed banking is not a viable option for many orchids, so a living population must be maintained to save them from extinction. As so many orchid-rich regions are simultaneously under threat of poaching, destruction, and climate change, many botanists and hobbyists have begun propagating endangered orchids *ex situ*, meaning outside of their endemic habitat. These orchids then are returned to the wild. But of course, the best plan of action to save orchids around the world is to preserve their natural environments.

Orchids are an excellent plant family to focus conservation efforts upon because they are indicators of species diversity and environmental health of any region more broadly. Lou Jost, an ecologist in Ecuador and president of Fundacion EcoMinga, an organization dedicated to the conservation of areas of Ecuador on the edge of the Amazon basin, explains that when we preserve orchids, most other species of plants, birds, and bees in an affected area benefit as well. Jost is a physicist and mathematician by training, but his time in the field with applied expertise has enabled

Lou Jost, *Phragmipedium lindenii* (2005).

him to publish useful advances in ecology, genetics, and conservation biology. Jost advises that one of the best methods of protecting orchids and their ecosystems is the purchase of swaths of biodiverse "hotspots" for conservation. The strategy is most effective where the topography is varied—more species are saved where an area has mountains, repeated changes in elevation, rivers, lakes, and a general diversity of landforms. International conservation charities like the Orchid Conservation Alliance, World Land Trust, and Rainforest Trust fund partners around the world like Jost's EcoMinga Foundation, who then buy and manage such preserves in orchid-rich areas of the world.

Realizing that orchids are important for human groups, too—and appreciating that local support is essential for the long-term success of any reserve—these programs work to incorporate the surrounding community in the care of the reserves. In EcoMinga's case, the Dracula Orchid Reserve in Ecuador's Choco region has worked to support a community group that invites reserve visitors and researchers to stay in their homes, for which the hosts receive payment for room and board. Visitors pay their hosts directly, and the program has had the side effect of empowering women in the community, who are in charge of the program. Foundations like EcoMinga create economic and educational benefits for the regions they protect, beyond any one preserve's boundaries. These places are not immune from climate change, of course, but they are defended from land conversion, mining, and poaching.

In his work in Ecuadoran cloud forests, Jost has added dozens of new orchids to the scientific record, including previously unnamed lepanthes, maxillaria, epidendrum, habenaria, teagueia, masdevallia, and species in several other genera. In addition to his science, he is also an artist, painting the orchids, birds, butterflies, fish, and ocelots he works to protect. And like Mee and von Foerster, his art has been an essential means of communicating the beauty and wonder of Amazonia. He explains, "A good photo or painting moves people far more than a set of facts and figures . . . [it forms] a deep connection between the artist and subject, and between the viewer and subject."

In 2020 alone, the world lost thirty million acres of tropical forest. Recent studies on the decline of native orchids have shown that this is not just an ecological problem, it is also a human problem—ongoing access to plants is important for intergenerational knowledge transmission, cultural identity, and human health around the world. The international web of botanical gardens, ecologists, artists, reserves, and Native peoples working toward preserving our most sensitive orchid flora, has proven to be one of the best models for maintaining the diversity of other endangered species. Although orchid preserves are relatively new in concept, several orchids previously unknown to science have already been discovered in protected areas, and the plant life so essential to global climate patterns and human imaginations are thriving.

SOBRALIA FRAGRANS

Sobralia margaritae is not widely used in the orchid trade, but there are many other sobralias to choose from. Their colors can range from white, yellow, and green to pink, purple, cerulean, and brown, and species may grow from one foot to over twenty-five feet tall. Most used as a garden plant in warm regions, gardeners appreciate its bamboo-like foliage and sweetly scented flowers. *Sobralia*

fragrans echoes *S. margaritae* in that it hails from Amazonia, smells delicious, and is a beautiful butter yellow, but is a petite orchid, better suited for collections. Here, booted racket-tail hummingbirds inspect S. *fragrans* in Ecuador in John Gould's *A Monograph of the Trochiliae* (1861). *Sobralia powelli* and *S. rosea* may also be good choices for today's growers, depending on your taste and conditions.

Orchid Details

— Place of origin: tropical Amazonia; Central and South America
— Blooming season: usually May through July; blooms whenever new canes mature throughout the year
— Flowers last for: usually only one day, but an inflorescence can bloom successively for several days
— Plant size: less than 12 inches high; will form a clump if allowed the room
— Flower size: 2.5–3 inches
— Fragrance: sweet, honey
— Plant habit: sympodial, stems upright, slightly clumping

Orchid Needs

— Light: medium shade; can take morning sun
— Humidity: 30%–80% and very good air movement
— Water: when in active growth, must be watered well; in cooler temperatures, growth slows, so water less
— Growing medium: fast draining—large bark with perlite if grown indoors; gritty substrate if grown outdoors; this sobralia grows epiphytically as well as terrestrially, so if your conditions permit, you can mount the orchid on bark or cork

⌒ Fertilizing schedule: fairly heavy feeders, they appreciate full-strength balanced fertilizer with every other watering

⌒ Seasonal changes: 50°F–90°F year round

⌒ Special requirements: sobralias resent being repotted; only do so when you see new growth and new root tips

Appendices

Glossary of Orchid Terms

❧

ALLIANCE An informal grouping of closely related orchid genera; for example, the cattleya alliance contains cattleya, brassavola, and laelia, among others.

ANTHER The part of the stamen that bears pollen in a flower.

COLUMN Fused male and female sexual organs (stamens, style, and stigma) in the center of an orchid flower, usually waxy and white and held above the labellum.

GENUS A group of closely related species with an assumed common

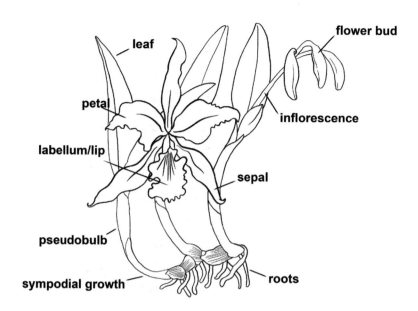

ancestry; one taxonomic rank above species. There are roughly 880 naturally occurring orchid genera, and roughly 500 additional human-made intergeneric ones. Plural: **genera**.

HYBRID A **primary hybrid** is the progeny of the union between two different species of orchids; a **complex hybrid** is a cross between a species and a hybrid, or two hybrids.

INFLORESCENCE The flower stem, buds, and flowers on an orchid. Colloquially called a **spike**.

LABELLUM A modified petal on an orchid, usually showy, and acts as a landing platform for pollinators. Also known as the **lip**.

MONOPODIAL One of two forms of an orchid in which the plant meristem (active growing point) grows vertically, for example, in pha-

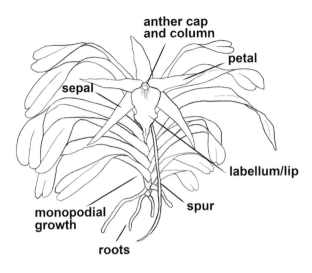

laenopsis, vanda, and angraecum. By contrast, **sympodial** growth in orchids arise from the base of the previous growth and generally grow laterally, for example, cattleya and cymbidium.

PETAL One of three inner parts of an orchid flower, positioned between three outer sepals. One petal is often modified into the labellum of the orchid.

POLLINIUM Waxy collection of pollen found in the anthers of most orchids; often yellow and found under the cap of the column. Plural: **pollinia**. Also known as: **pollen packet**.

PSEUDOBULB Thickened plant tissue at the base of a leaf serving as nutrient storage for a sympodial orchid.

ROSTELLUM Interior plant tissue of an orchid projecting from the column, separating anther from stigma, preventing self-pollination.

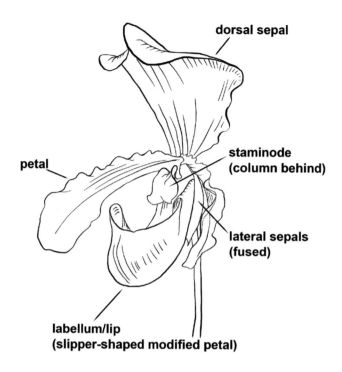

dorsal sepal

petal

**staminode
(column behind)**

**lateral sepals
(fused)**

**labellum/lip
(slipper-shaped modified petal)**

SCIENTIFIC NAME The scientific name of an orchid contains the genus and the species, for example, *Angraecum sesquipedale*. Orchid growers also add variety and/or cultivar names, for example, *Guarianthe bowringiana var. coerulea* or *Ansellia Africana* 'John Hope Franklin.'

SEPAL One of three outer segments of an orchid flower, positioned around three petals. The **dorsal sepal** is topmost (for example, on a paphiopedilum); **lateral sepals** are lower.

SPECIES The basic unit of biological classification; a group of highly similar plants. There are over twenty-six thousand species of orchids. **Type species** is the species within a genus that is thought to be the best example of the characteristics of the genus, for example, *Cattleya labi-*

ata, Dendrobium moniliforme, Paphiopedilum insigne, and *Phalaenopsis amabilis* are all type species for their respective genera.

SPUR A tubular, hollow extension of the base of some orchids (usually the lip of the orchid), sometimes containing nectar for pollinators.

STAMINODE An infertile stamen that covers the stigma and anthers of some orchids; the "button" at the center of a paphiopedilum orchid.

Top Fifteen Tips for Growing Orchids

❧

1. **JOIN LOCAL AND NATIONAL ORCHID SOCIETIES.** The guidance you receive from growers active in your region is far more valuable than any glib advice found online. Orchid societies offer expert lectures, webinars, group trips, and deals on orchids and orchid supplies. But most people report that the real benefit of joining a society is the friends they make. So, in addition to being a better grower, you'll pick up kindred spirits, too.

2. **MATCH ORCHIDS TO YOUR MICROENVIRONMENTS.** Do you have sunny or shady windowsills? A well-lit mudroom that cools down in the winter? A screened four-season porch? Although most

orchid fanatics buy pretty orchids and ask questions later, picking orchids that will suit the environment you can provide for them will prevent inevitable frustration with ailing plants.

3. **INSPECT AND QUARANTINE ORCHIDS UPON PURCHASE.** Like all plants, orchids can come with hitchhikers—mealybugs, thrips, scale, snails, and the like. Upon purchase, if possible, pop the orchid out of its pot and inspect its growing media, roots, pseudobulbs, leaves, and flowers for bugs and decay. If the orchid is in active growth, consider repotting it with fresh media to make sure all hangers-on are removed. And even if your orchid looks pristine, it's always a good idea to keep it away from the rest of your collection for at least six weeks—the incubation period for many bugs and diseases.

4. **MONITOR YOUR POTTING MEDIA.** No mold allowed! Do not allow your orchid to grow white mold ("snow mold") on its roots or in its potting media. The mold is a sign that the potting media is rotting and probably too acidic for the orchid. Do your best to change the potting media as soon as possible. Soak the pot in luke-warm water to loosen the media, toss the spent media in your com-post bin, and repot with fresh media. Remove dead (brown and soft) roots from the orchid, and spray any roots that had the mold on them with hydrogen peroxide before repotting. Most orchids need to be repotted with fresh media every one to two years.

5. **USE THE SMALLEST POSSIBLE POT.** Orchid roots much prefer to be crowded rather than swimming (and rotting) in a pot that is too large for them. Sometimes small pots and large leaves can lead to a top-heavy plant—address that by placing the orchid's small pot inside of a heavier clay or decorative pot. (This also helps to further regulate the humidity around the roots.) Small pots for the

orchid's size—even if it means that some of the roots crawl out of the pot—in the end make for happier orchids.

6. **REMEMBER THAT MANY ORCHIDS POPULAR TODAY ARE EPIPHYTES.** They evolved to grow on trees, with moving air, picking up nutrients from the rain, with little media surrounding their roots. Very few orchids like to sit in water, and most like to be "just dry" before watering them again. This is why it is impossible to definitively answer the question of how often to water one's orchid—each orchid in each potting media can vary.

7. **FLUSH POTS REGULARLY WITH PURE WATER.** Most orchids love a good shower. If grown indoors, regularly take your orchid to the sink and spray the leaves, pseudobulbs, media, and roots. This prevents the buildup of dust, washes away mites and other critters, and flushes the pot of accumulated fertilizer salts. Many growers recommend doing this before you fertilize the orchid, every week to month or so.

8. **PURCHASE A GOOD-QUALITY, BALANCED ORCHID FERTILIZER.** There are a bewildering array of orchid fertilizers available today. Over the past decade, most enthusiasts have settled on using a balanced fertilizer with micronutrients—like the "MSU Formula"—with occasional amendments as needed. Michigan State University's fertilizer is now widely available and in addition to nitrogen, potassium, and phosphorus (NPK), includes essential "micronutrients" calcium and magnesium. It can be used across a variety of genera, but be sure to follow specific advice for each orchid, rather than indiscriminately fertilizing all orchids all year. If you own several orchid genera, you may want to set up a fertilizing schedule for yourself for each month of the year. In addition to MSU, some orchids like an extra boost of phosphorus, calcium, and magnesium in peak growing season.

9. **KEEP HUMIDITY ABOVE 40%.** Plants breathe through their leaves and roots; most orchids need significant moisture in the air to stay healthy. If you keep them in a spot where the humidity is reliably 40% or above, they will naturally fare better and mites and other bugs won't reproduce as quickly. Beginners often benefit by placing their orchids on a shaded windowsill in the bathroom or kitchen—the plants love the humidity and the ambient temperature spikes in those rooms. Humidity monitors are inexpensive, usually include temperature gauges, and are handy in any growing space.

10. **EVALUATE ORCHID LIGHTING.** Orchids are classed into three light levels: low (quite shady for most phalaenopsis and paphiopedilums), medium (bright shade for oncidiums, dendrobiums, and cattleyas), and high (some direct sun for many vandas and cymbidiums). There are many ways of estimating light levels in your home, but I prefer to remove the guesswork and use a light meter—available through several apps for smartphones. If you supplement your orchids' light with artificial lighting, full-spectrum, energy-efficient LED lights are highly recommended. But the single best way to judge whether your orchid lighting is appropriate is by looking at your orchid: If the leaves are grassy green, and the plant is growing normally and blooming regularly, all is well. Dark green leaves indicate light levels are too low; yellowing or reddening of the foliage and stunted growth can point to light being too high.

11. **LEARN ABOUT POTTING MEDIA, AND THEN EXPERIMENT.** There are two types of media that growers use most for phalaenopsis orchids and many other genera: bark and long-fiber sphagnum moss. But there are many other types of media to try—chunky perlite, hydroton (LECA), seramis, charcoal, tree fern, and the list goes on. Unfortunately, many bagged "orchid mixes" sold

at big box stores are pretty terrible for orchids—the bark is too small, old, and dry. Your best bet is to visit a local garden center or orchid greenhouse and inquire about what they might recommend, buy good-quality bark and/or long-fiber sphagnum moss, or a genera-specific mix from a specialty grower. Like many folks in my local orchid society, after decades of growing orchids, I'm still experimenting with media. Many orchids bought today come in jam-packed pots of sphagnum—this is not good for the orchid, especially in home environments. If you want to keep your orchids in pure sphagnum moss, remove the orchid and moss from the pot, soak it for half an hour, remove the moss from the orchid, drain it, and repack the moss around the orchid in the pot *very* loosely. This ensures that enough air is getting to the roots, preventing rot and overcrowding.

12. **TRY MOUNTING ORCHIDS.** For many years I only grew orchids in pots—I was afraid of mounting orchids on cork or bark because I couldn't fathom them receiving the water and nutrients they needed and feared making a mess in my limited orchid space. But thanks to an adventurous friend, we tried it together, and now in addition to mounting any orchid that's a petite epiphyte, I mount any orchid that's ailing in my collection. When in doubt, more air around an orchid's root ball usually solves the problem. Mounting orchids often makes them even more beautiful, is fun, and saves space—you can now grow vertically as well as horizontally.

13. **STERILIZE TOOLS AND POTS.** Orchids can carry visible and unseen viruses, including Odontoglossum ringspot virus (ORSV), Cymbidium mosaic virus (CyMV), the orchid fleck virus (OFV), and pathogens like fusarium. Some growers invest hundreds to thousands of dollars per year testing their orchids for viruses. Whether or not you go that far, excellent tool and pot sanitation

is crucial in keeping your orchid collection safe from widespread infection. (And *never* share used water between orchids.) Viruses are most quickly spread when orchids with open wounds touch other orchids with open wounds—or their "sap" is carried on a tool or in a pot. The best way to sanitize your tools is to set the blade in a blue flame for several seconds, and then soak it in a 10% bleach solution. If you do reuse your pots, scrub them of debris and soak in a 10% bleach solution for a few hours.

14. **ORGANIZE AND MAINTAIN ORCHID RECORDS.** If you find your orchid collection is growing, it's wise to keep records of when and where each plant was purchased, when it last bloomed, when it was last repotted, whether you've tested it for viruses, and if you have given divisions to friends. Some growers keep all of this information written on tags kept within the orchid pot. But once your collection reaches a certain size, create a spreadsheet for long-term record keeping.

15. **CREATE AN ORCHID OASIS.** Most orchids grow best in moderate to high humidity, full-spectrum light, and with a 10°F–15°F drop in day-to-night temperatures. For many avid in-home growers, creating a dedicated space that provides orchids with their ideal conditions doesn't just benefit the plants; it provides a green retreat for their wranglers too. This might be in the form of a small orchidarium, a converted cabinet, a rolling chef's rack, or a basement grow tent. Make sure your space has seating room—you'll want to be there, breathe there, and enjoy it as much as possible.

Acknowledgments

✣

M y orchid journey has been archival, personal, and enriched by a diverse community of flower fanatics and cultural history lovers. I would not have been able to write this book without the A.D. and Mary Elizabeth Andersen Hulings Distinguished Chair in the Humanities at Northland College, a multiyear sabbatical program designed to promote original work in the humanities. In some ways, this book can also be sourced back to the US National Gallery of Art Alisa Mellon Bruce Predoctoral Fellowship for Historians of American Art to Travel Abroad I held more than a decade ago. It allowed me to perform a glorious summer research tour of botanical gardens in England, Sweden, and France, and inspired me to think much more broadly about the role of plants and botanical gardens in art and world history.

A special thanks goes to my intrepid colleagues at Northland College, especially those who read chapter drafts for me, including Les Alldritt, Paul Schue, and Brian Tochterman. Sarah Johnson, thank you for my first real lessons in botany, my first loupe, and for glorious days in the bogs and fens of Northern Wisconsin hunting for native orchids. Several other Northland colleagues offered enthusiastic moral support and/or delivered generative resources when I inquired of them, including Cynthia Belmont, Kyle Bladow, Charles Krysinski, Emily Macgillivray, Danny Simpson, and Angela Stroud.

My wonderful colleagues and friends elsewhere, including Paul Bogard, Theresa Kelley, Alex Mendes, and Pam Steinle, varyingly read the book proposal and chapter drafts, discussed the framing of this book, or even helped collect resources from afar during the COVID-19 pandemic. This book, too, would not have been as rich—nor as satisfying to write—without the help of several thoughtful, generous, and brilliant interviewees. Thanks for sharing their time and expertise goes to Cathy Davidson, John Whittington Franklin, Lou Jost, and Madeline von Foerster.

I am endlessly thankful for several archives and libraries and the people that direct and work in them. Doug Holland, director of the Raven Library at the Missouri Botanical Garden, opened the door to archival botanical research for me long before this book was conceptualized, and remains a thoughtful friend. My unending thanks go to Northland College's Dexter Library staff, especially our interlibrary loan sleuth extraordinaire Elizabeth Madsen-Genzler (she even does home delivery!) and library director Julia Waggoner. Thanks also to Rob Strauss, special collections and archives coordinator at Cal Poly Pomona, who tracked down details of Raymond Burr's orchids. And in relative order of the in-person and online archives required to write this book, my gratitude to an abbreviated list of institutions includes the Biodiversity Heritage Library; wider Smithsonian Libraries; the University of Wisconsin, Madison; the University of Minnesota, Twin Cities; the Library of Congress; the Royal Botanic Gardens at Kew; and the British Museum.

Now to thank my orchid-loving tribe! The Northland Orchid Society in Duluth, Minnesota, and the Orchid Society of Minnesota in Minneapolis have astounding—and astoundingly friendly—orchid growers. I learn constantly from all of them. Special thanks go to Ross Hella, who helped me through several technical horticultural and orchid breeding questions and navigated the depths of the OrchidWiz program. His skill as a paphiopedilum breeder is matched by his wit and friendship. Julie Hella, too, is a wonderful welcoming presence and a gifted orchid grower. And to everyone else in the Northland Orchid Society—especially Elourine Alspach, Gary Anderson, Julie Calligure, Orville Fay, Walter Lehenbauer, Charles Mans, and Anna and Bill Morrison—thanks for distracting me with orchids through long winters in the great white north! And since my time in the larger Orchid Society of Minnesota, the membership has been expertly, and lovingly, led by a large group a talented and hilarious people. My special thanks go to Roy Close, Michael Dyda, Steve Fillmore, Esteban Gonzalez, Melissa Merchant, Linda Smith, Steve Ukasick, and Katherine Weitz. I would not have grown nearly as fast an orchid fanatic without each of you. I'd also like to thank the American Orchid Society leadership, including *Orchids* editor Jean Allen-Ikeson and past president Ron McHatton for their help in tracking down sources for this book. Orchid world superstars including Bruce Rogers, Arthur Chadwick, Hermann Pigors, Elbert Wijaya, Charles and Susan Wilson, and Brenda Oviatt receive my deep thanks for their support, answers to several inquiries, and for thoughtful and thorough reviews of orchid care and science in the manuscript. (Any remaining errors are of course my own.)

Writing this book has been nothing less than dreamy. My keenest thanks go to my agent, Deirdre Mullane, whose time, support, and enthusiasm for this project is one of the greatest gifts I've ever received. And sincere gratitude for my editor, Amy Cherry, whose edits, clarifications, and requests for deeper life histories of the characters here—as well as her keen eye for style—made this a far better book. Special thanks to Huneeya Siddiqui, editorial assistant, whose patience, attention to detail, and good humor are constant resources for me. And Steve Colca, Julia Druskin, Charlotte Kelchner, Victoria Keown-Boyd, Yang Kim, Elizabeth Riley, and Oliver Wearing helped bring this book to life, gave it a high polish, and made sure plant lovers get their hands on it.

More personally, I also offer gratitude to my parents, Mike and Kari Hannickel, for educating their grandson for the first year of the pandemic, thereby allowing me to continue work on this book.

Finally, I'd also like to thank my friends in all things (especially flowers, gardens, and nineteenth-century dramas) Gayle Chatfield, Jachyn Davis, Kyra Flynn, Amber Pickney, Traci Sinclair, and Talia Starkey. Thank you for your spirited support, for being enablers of my plant habit, and for being your fabulous selves.

And most of all, to Jason and Miles Terry, who put up with too many orchids in our home and endless trips to botanical gardens and greenhouses, and whose love has utterly altered and radically enriched my life. I love you.

Notes

✿

Ch. 1 Lusty Ladies of the Enlightenment

6 **"Fair CYPREPEDIA with successful guile":** Erasmus Darwin, *The Botanic Garden, A Poem in Two Parts. Pt. 1 Containing the Economy of Vegetation, Pt. 2 The Loves of the Plants. With Philosophical Notes* (London: J. Johnson, 1791), 204–5.

8 **"it may be easily mistaken":** Darwin, *Botanic Garden*, 26.

9 **"chef d'oeuvre, the masterpiece":** Darwin, *Phytologia, or, the Philosophy of Agriculture and Gardening* (London: J. Johnson, 1800), 103.

9 **"Hence on green leaves":** Darwin, *The Temple of Nature* (London: J. Johnson, 1802), 63.

9 **"could never forsake the charms":** Janet Browne quotes an unnamed obit-

uarist in "Botany for Gentlemen: Erasmus Darwin and *The Loves of the Plants*," *Isis* 80, no. 4 (1989): 593–621.

10 **"did by art poetic transmute":** Darwin, *Botanic Garden*, vii–viii.

10 **"seized such hold of my":** All quotes in this paragraph found in Alan Bewell, "Erasmus Darwin's Cosmopolitan Nature," *ELH* 76, no. 1 (Spring 2009): 19–48.

12 **"an air of voluptuousness":** Richard Polwhele, *Unsex'd Females* (London: Cadell and Davies, 1798): author's note 21.

12 **"Cupid's cave," "Cupid's hotel,":** Phrases and words in this paragraph found throughout Peter Fryer, *Mrs. Grundy: Studies in English Prudery* (New York: London House & Maxwell, 1963).

13 **"I hardly do them any":** Quoted in Susan Branson, "Flora and Femininity: Gender and Botany in Early America," *Commonplace: The Journal of Early American Life*, accessed November 22, 2021, http://commonplace.online/ article/flora-femininity/.

14 **"With blushes bright as morn":** Darwin, *Botanic Garden*, 112–13.

15 **"insipid mucilaginous taste":** Darwin, *Botanic Garden*, 112–13.

16 **"Two Harlot-Nymphs . . . incommode them all":** Darwin, *Botanic Garden*, 117–18.

Ch. 2 Orchids Fit for a Chinese Empress

24 **"discreet yet interesting":** Maggie Keswick, *The Chinese Garden: History, Art, and Architecture* (Cambridge, MA: Harvard University Press, 2003), 191–92. See also Richard Barnhart, *Peach Blossom Spring: Gardens and Flowers in Chinese Paintings* (New York: Metropolitan Museum of Art, 1983), 55.

26 **"I have often thought that":** quoted in Princess Der Ling, First Lady in Waiting to the Empress Dowager. *Two Years in the Forbidden City* (New York: Moffat, Yard, 1911), 356.

27 **"flowers everywhere!":** Katherine Carl, *With the Empress Dowager of China* (New York: Century, 1905), 36.

30 **"walk[ed] along the long veranda":** Princess Der Ling, *Two Years in the Forbidden City*, 87.

30 **"every now and then potted":** Jung Chang, *Empress Dowager Cixi: The Concubine Who Launched Modern China* (New York: Anchor, 2013), 314.

30 **"Out of the Forbidden City":** Sterling Seagrave, *Dragon Lady: The Life and Legend of the Last Empress of China* (New York: Vintage, 1992), 5.

32 **"Leaves are painted in a":** Chieh Tzu Yuan Hua Chuan, *The Mustard Seed Garden Manual of Painting* (1679), translation edited by Mai-Mai Sze (Princeton, NJ: Princeton University Press, 1992), 325, 327.

34 **"The fire, fueled by more":** Chang, *Empress Dowager Cixi*, 33.

34 **"When we first entered":** Chang, *Empress Dowager Cixi*.

34 **"When our motherland was weak":** Qian Xingjian, *Famous Flowers in China* (Shanghai: Shanghai Press, 2010), 31–33.

34 **"She had flowers always about":** Katharine Carl, *With the Empress Dowager of China*, 40.

Ch. 3 Orchids in the Tenderloin

40 **By the 1890s, a night:** Lucy Sante, *Low Life: Lures and Snares of Old New York* (New York: Farrar, Straus, and Giroux, 1991), 115.

41 **"waxlike human faces":** "Wonders of the Orchids," *New York Times*, March 2, 1887, 2.

41 **"red and bright colored orchids":** "Orchids in Full Bloom," *New York Times*, February 17, 1888, 2.

41 **"confirmed monomaniacs on orchids":** "Wonders of the Orchids," *New York Times*, March 2, 1887, 2.

42 **"gleamed like green polished moccasins":** "Wonders of the Orchids," *New York Times*, March 2, 1887, 2.

43 **"saucy-looking lavender-eyed":** "Orchids in Full Bloom," *New York Times*, February 17, 1888, 2.

44 **In the gallery above:** Details in this paragraph found in *Eden Musée Monthly Catalog, January 1892* (New York: Rich G. Hollman, 1892).

44 **"almost believe himself in fairy-land":** "Orchids in Full Bloom," *New York Times*, February 17, 1888, 2; see also Emily Louise Taplin, "The New York Orchid Show," *The American Florist*, March 1, 1888, 319–21.

44 **The museum made a special:** "Orchids in Full Bloom," *New York Times*, February 17, 1888, 2; "A Wealth of Flowers: Auspicious Opening of the Show at the Eden Musée," *New York Times*, November 21, 1888; "Beauty in Flowers," *New York Times*, February 28, 1889, 8; "The Orchid Show," *New York Times*, March 3, 1892, 9.

46 **"Orchids of the Future":** "Orchids of the Future," *The American Florist*, April 15, 1887, 349.

46 **"orchids are like diamonds"**: "Among Rare Orchids," *New York Times*, February 20, 1890, 8.

49 **number of orchids exhibited multiplied:** "The Fifth Orchid Show: No Expense to be Spared in Making It Successful," *New York Times*, February 21, 1889, 5.

50 **"with as much care"**: "Orchids in Full Bloom," *New York Times*, February 17, 1888, 2.

50 **"monotony is an unknown quantity"**: "The Orchid Show," *New York Times*, March 3, 1892, 9.

Ch. 4 Frida Kahlo's Orchid

59 **"lived dying"**: quoted in Gannit Ankori, *Imagining Her Selves: Frida Kahlo's Poetics of Identity and Fragmentation* (Praeger, 2002), 101.

62 **"evocative of wild and distant"**: Ruth Mosher Place, "Thrilling as a Jungle Movie: Growing Orchids in Detroit," *Detroit News*, March 26, 1939.

63 **"foolishly vulgar . . . Wife of the Master"**: Editorials regarding Rivera's murals appeared in the *Detroit News* on March 18, 1933, and March 21, 1933; Florence Davies, "Wife of the Master Mural Painter Gleefully Dabbles in Works of Art," *Detroit News*, February 2, 1933.

65 **"had the idea of a sexual"**: Quoted in Hayden Herrera, *Frida Kahlo: The Paintings* (New York: HarperCollins, 1991), 72.

65 **"nothing, to the feminine taste"**: Ruth Mosher Place, "Thrilling as a Jungle Movie: Growing Orchids in Detroit," *Detroit News,* March 26, 1939.

66 **"mythical creature"**: Hayden Herrera, *Frida: A Biography of Frida Kahlo* (New York: Perennial, 2002), 116.

69 **"had the daintiness of miniatures"**: Quoted in Phyllis Tuchman, "Frida Kahlo," *Smithsonian Magazine* (November 2002), https://www.smithsonianmag.com/arts-culture/frida-kahlo-70745811/.

Ch. 5 Rafinesque's Strange Collections

74 **"gloom of solitary forests"**: C. S. Rafinesque, *New Flora and Botany of North America* (Philadelphia: H. Probasco, 1836), 12.

74 **"rough or muddy roads"**: Rafinesque, *New Flora and Botany of North America.*

75 **"he is doubtless a man":** Quoted in Ronald L. Stuckley, "Opinions of Rafinesque Expressed by His American Botanical Contemporaries," 31.

76 **"thought [Rafinesque] had gone mad":** Quoted in Alison Flood, "John James Audubon and the Natural History of a Hoax," *The Guardian*, May 3, 2016, https://www.theguardian.com/books/booksblog/2016/may/03/john-james-audubon-and-the-natural-history-of-a-hoax.

78 **the yellow lady's slipper bloomed:** C. S. Rafinesque, *Medical Flora of the United States* (Philadelphia: Atkinson and Alexander, 1828), 144.

80 **"all kinds of adventures, fares":** Rafinesque, *New Flora and Botany of North America*, 11.

80 **"Every pure botanist":** Rafinesque, *New Flora and Botany of North America*, 15.

81 **Rafinesque's greatest "misfortune":** Quoted in Leonard Warren, *Constantine Samuel Rafinesque: A Voice in the American Wilderness* (Lexington: University Press of Kentucky, 2004), 205; see also "A Sketch of the History of Conchology in the United States," *The American Journal of Science* 98 (March 1862) Second Series, 163.

83 **"the Japanese, Chinese, Hindu":** C. S. Rafinesque, *Flora Telluriana* (Philadelphia: H. Probasco, 1836–1838), 38–39.

85 **"beautiful and modest woman":** C. S. Rafinesque, "Lecture on Knowledge," unpublished manuscript, presented to the American Philosophical Society, November 7, 1820. Quoted in Charles Boewe, "Rafinesque Among the Field Naturalists," *Bartonia* 54 (1988), 49.

85 **"I hope to become":** Rafinesque, *Flora Telluriana*, 7.

Ch. 6 The Wind Orchid

88 **"You can get off alcohol":** Quoted in Eric Hansen, *Orchid Fever: A Horticultural Tale of Love, Lust, and Lunacy* (New York: Vintage, 2000), ii.

94 **"diligent, intelligent, and unassuming":** Quoted in Timon Screech, *Japan Extolled and Decried: Carl Peter Thunberg's Travels in Japan, 1775–1776* (New York: Routledge, 2005), 5.

95 **"they brought to me":** Preface to the first sections of Thunberg's *Flora Japonica*, 1784, translated and compiled in Timon Screech, *Japan Extolled and Decried: Carl Peter Thunberg's Travels in Japan, 1775–1776* (Routledge, 2005), 255.

95 **"It afforded me less pleasure"**: Quoted in Timon Screech, *Japan Extolled and Decried*, 175.

97 **"Every year there is increase"**: Charles Peter Thunberg, "Botanical Observations on the Flora Japonica" (read to the Linnaean Society of London, October 1, 1793), reprinted in *Transactions of the Linnaean Society of London* (London, 1794), 2:326.

97 **He happily regaled his audience**: Thunberg, "Botanical Observations on the Flora Japonica."

99 **"strange, peculiar species"**: "Oceoclades falcata," *Gartenflora* 15 (1866), 69.

Ch. 7 The Science of Freedom and Charles Darwin's "Little Book on Orchids"

105 **"I am a gambler, & love"**: Charles Darwin to J. D. Hooker, 26 March 1863, Darwin Correspondence Project, "Letter no. 4061," accessed September 1, 2021, https://www.darwinproject.ac.uk/letter/DCP-LETT-4061.xml.

106 **"universally acknowledged to rank amongst"**: Charles Darwin, *The Various Contrivances by Which Orchids are Fertilized by Insects* (New York: D. Appleton, 1877), 1–2.

106 **"multiform & truly wonderful"**: Charles Darwin to George Gordon, September 17, 1860, Darwin Correspondence Project, "Letter no. 2920," accessed September 1, 2021, https://www.darwinproject.ac.uk/letter/DCP-LETT-2920.xml.

106 **"At last gleams of light"**: Darwin to J. D. Hooker, January 11, 1844, Darwin Correspondence Project, "Letter no. 729," accessed September 2, 2021, https://www.darwinproject.ac.uk/letter/DCP-LETT-729.xml.

107 **"moths and butterflies perform their"**: Darwin, *Various Contrivances*, 15, 34.

108 **"Botanical ignoramus"**: Darwin to Hooker, December 25, 1844, Darwin Correspondence Project, "Letter no. 803," accessed September 14, 2021, https://www.darwinproject.ac.uk/letter/DCP-LETT-803.xml.

109 **"It is interesting to contemplate"**: Charles Darwin, *On the Origin of Species* (Cambridge University Press, 2009 [first published 1859]), 376.

110 **Even Darwin's children reminisced**: Francis Darwin, *The Life and Letters of Charles Darwin, Including an Autobiographical Chapter* (London: John Murray, 1887), 1:116; *Emma Darwin, Wife of Charles Darwin. A Century of Family Letters* (Cambridge University Press, 1904), 2:376–77.

110 **"An examination of [orchids'] many":** *Various Contrivances*, 2.

110 **the comet orchid, the king:** Costa, *Darwin's Backyard*, 255.

113 **In January of 1862:** Darwin to J. D. Hooker, January 25, 1862, Darwin Correspondence Project, "Letter no. 3411"; and Darwin to J. D. Hooker, January 30, 1862, Darwin Correspondence Project, "Letter no. 3421," https://www.darwinproject.ac.uk/letter/DCP-LETT-3411.xml.

114 **"If you can really spare":** Darwin to J. D. Hooker, November 1, 1861, Darwin Correspondence Project, "Letter no. 3305," accessed September 2, 2021, https://www.darwinproject.ac.uk/letter/DCP-LETT-3305.xml.

114 **"Do you really think I":** Darwin to T. H. Farrer, June 5, 1868, Darwin Correspondence Project, "Letter no. 6230," accessed September 2, 2021, https://www.darwinproject.ac.uk/letter/DCP-LETT-6230.xml.

114 **"appears to me one of":** Darwin, *Various Contrivances*, 44.

114 **"such cases shake to the":** John Lindley, *The Vegetable Kingdom: The Structure, Classification, and Uses of Plants*, 3rd edition (London: Bradbury & Evans, 1853), 178. Darwin references this in *Various Contrivances*, 196.

114 **"In my examination of Orchids":** Darwin, *Various Contrivances*, 284.

115 **"It is hardly an exaggeration":** Darwin, *Various Contrivances*, 293.

115 **"How then does Nature act?":** Darwin, *Various Contrivances*, 179.

115 **"The study of these wonderful":** Darwin, *Various Contrivances*, 224–25.

117 **"that the North would proclaim":** Darwin to Asa Gray, June 5, 1861, Darwin Correspondence Project, "Letter no. 3176," accessed September 13, 2021, https://www.darwinproject.ac.uk/letter/DCP-LETT-3176.xml.

117 **"I am very poorly today":** Darwin to Charles Lyell, October 1, 1861. Darwin Correspondence Project, "Letter no. 3272," accessed September 9, 2021, https://www.darwinproject.ac.uk/letter/DCP-LETT-3272.xml.

117 **Asa Gray wrote to Darwin:** Asa Gray to Charles Darwin, Darwin Correspondence Database, Entry 3637 (July 2, 1862) and Darwin to Gray, Entry 3662 (July 23–24, 1862), https://www.darwinproject.ac.uk/letter/DCP-LETT-3662.xml.

Ch. 8 Itinerant Orchids, Enslaved People

131 **"This clever boy had realized":** Bellier-Beaumont letter to Sainte-Suzanne's Justice of the Peace, Monsieur Ganne, in 1861, quoted in Tim Ecott, *Vanilla: Travels in Search of the Ice Cream Orchid* (New York: Grove Press, 2004), 141.

133 **A newspaper in Réunion recorded:** Quoted in Ecott, *Vanilla*, 147.

133 **Albius's former master appealed:** Quoted in Ecott, *Vanilla*, 139–41.

134 **"veritable magic wallet":** Joseph Burnett, *About Vanilla* (Boston: Joseph Burnett, 1900)

Ch. 9 Jane Loudon and Her Floriferous Press

143 **"our chances of being happy":** Mrs. Loudon, *Lady's Country Companion, or, How to Enjoy a Country Life Rationally* (London: Longman, Brown, Green, and Longmans, 1845), 7.

143 **"I like to see a lady":** Loudon, *Lady's Country Companion*, 251.

144 **In a time of very poor:** Loudon, *Lady's Country Companion*, 162–70.

144 **Her teenage pen was full:** Jane Wells Webb, *Prose and Verse* (London: R. Wrightson, 1824), 81, 105.

147 **For Jane's virtuous Queen Elvira:** Anonymous [Jane Wells Webb], *The Mummy! Or a Tale of the Twenty-Second Century* (London: Henry Colburn, 1827; reprinted by Good Press, 2019), 290, 465.

148 **Jane both wrote about her:** Mrs. Loudon, *Lady's Country Companion*, 160, 128.

148 **Speaking to middle-class novices:** Loudon, *Lady's Country Companion*, 162–70.

149 **"[Loudon's] old woman":** George Glenny, *Horticultural Journal and Florists' Register* 1 (1834): 54; also quoted in Ray Desmond, "Victorian Gardening Magazines," *Garden History* 5, no. 3 (Winter 1977): 60.

150 **"is one of innocence and":** Anonymous [Jane Loudon], *The Young Lady's Book of Botany: A Popular Introduction to that Delightful Science* (London: Robert Tyas, 1838), 1.

151 **"the formation of the flowers":** Anonymous [Loudon], *Young Lady's Book of Botany*, 247.

151 **advice that remains patent:** Mrs. Loudon, *Instructions in Gardening for Ladies* (London: John Murray, 1840), 350, 552–53.

151 **"The orchidaceae are no less":** Anonymous [Loudon], *Young Lady's Book of Botany*, 247–48.

151 **"distinct and remarkable . . . lip":** Anonymous [Loudon], *Young Lady's Book of Botany*, 246.

152 **"Nothing is more natural":** Mrs. Loudon, *Botany for Ladies, or, a Popular Introduction to the Natural System of Plants* (London: John Murray, 1842), 1.

152 **"The beautifully colored parts"**: Loudon, *Botany for Ladies*, 4.

152 **"the ovary is juicy"**: Loudon, *Botany for Ladies*, 5–6.

152 **"still more varied and fantastic"**: Loudon, *Botany for Ladies*, 441.

152 **"We know that we ourselves"**: Loudon, *Botany for Ladies*, 7.

154 **"one of the handsomest"**: Mrs. Loudon, *British Wild Flowers* (William Smith, 1846, originally published 1844), 286.

154 **"there is perhaps none"**: Loudon, *British Wild Flowers*, 286.

154 **"a number of fancifully dressed"**: Loudon, *British Wild Flowers*, 289.

156 **helmut flower, *Coryanthes macrantha***: Mrs. Loudon, *Ladies' Companion to the Flower Garden* (1841, several editions), 73.

156 **"varies considerably in the different"**: Loudon, *Ladies' Companion to the Flower Garden*, 190.

157 **"Splendid orchideous epiphytes"**: Loudon, *Ladies' Companion to the Flower Garden*, 91.

158 **"No one, in fact, can"**: Loudon, *Ladies' Companion to the Flower Garden*, 371.

Ch. 10 Orchids and Steel

161 **On Sunday, February 17:** "The Orchid Craze at Its Height in Fashionable New York," *New York Times*, February 17, 1907, 2.

162 **The extended Vanderbilt family:** "All Society in Costume: Mrs. W. K. Vanderbilt's Great Fancy Dress Ball," *New York Times*, March 27, 1883, 1; "Vanderbilt-French Wedding: Floral Decorations to Be More Elaborate than First Intended," *New York Times*, January 10, 1901, 7; "Orchid Craze at Its Height in Fashionable New York," *New York Times*, February 16, 1907, 2; "Vanderbilt Wedding Plans Completed: House to Be Decorated Throughout with Orchids in All Colors," *New York Times*, January 26, 1908, 9.

163 **"many exquisite and rare varieties"**: "Jay Gould's Orchids," *Placer Herald* 44, no. 45 (from the *New York Herald*), September 26, 1896, 3.

164 **She was a patron:** Eve M. Kahn, "Tiffany Show Reveals Helen Gould's Role as Arts Patron," *New York Times*, June 12, 2018.

164 **"one of the finest collections"**: "Orchids the Favorite," *The Journal* (New York), March 28, 1896, 11.

165 **As the president:** Hamilton Schuyler, *The Roeblings: A Century of Engineers, Bridge-Builders and Industrialists* (New York: AMS Press, 1931), especially chapters 17–21; see also David McCullough, *The Great Bridge: The Epic Story of the Building of the Brooklyn Bridge* (New York: Simon & Schuster, 2001).

166 **"not only be the greatest":** John A. Roebling, Report to the New York Bridge Company, September 1, 1867.

167 **First Golden Age of paphiopedilum:** Highlights of slipper orchid history are found in Harold Koopowitz, *Tropical Slipper Orchids: Paphiopedilum and Phragmipedium Orchids* (Portland, OR: Timber Press, 2008).

169 **By 1895, Roebling had:** *The Orchid Review* 2 (1894), 324–25.

173 **"it was impossible for anyone":** *Transactions of the Massachusetts Horticultural Society for the Year 1910* (Boston 1911), 132–38. Plate 3 between pages 132 and 133 has picture of display.

173 **Roebling's major contributions:** *The Orchid Review* (October 1903), 311–12; *Gardeners' Chronicle* (1903, ii, p. 227, fig. 93); *The Florists' Exchange* 46 (1918), 585.

174 **Roebling often commemorated US presidents:** Oakes Ames records the name of Clinkaberry's *Paphiopedilum* James Garfield in *American Gardening* 21, no. 278 (April 21, 1900), 277.

175 **"one of the greatest pleasures":** *The Orchid Review* 27 (January–February 1919), 33–34.

176 **"the flower of the moment":** "Orchids the Favorite" *The Journal* (New York), March 28, 1896, 11.

Ch. 11 The Flowers, Fashion, and Friendships of Empress Eugénie

181 **"I was born during an":** Augustin Filon, *Recollections of the Empress Eugénie* (London: Castle, 1920), 8.

182 **"good fairy who was":** Washington Irving, *Tales of the Alhambra* (Philadelphia: Lea & Carey, 1832), 32.

183 **"How beauteous is this garden":** Irving, *Tales of the Alhambra*, 65.

185 **"beautiful, clever, very coquette":** Quoted in Desmond Seward, *Eugénie: The Empress and Her Empire*, loc. 1678, Kindle.

186 **"both in quality and extent":** "Grand Floral Fete at the Crystal Palace," *The Scotsman* (June 6, 1855), 2.

187 **As Émile De Puydt's lushly:** Émile De Puydt, *Les Orchidées, Histoire Iconographique, Organographie, Classification, Géographie, Collections, Commerce, Emploi, Culture, avec une Revue Descriptive des Espèces Cultivées en Europe* (1880), 4, 16, 160–61; translation from the French is mine.

190 **"like a fairy queen":** Quoted in Seward, *Eugénie: The Empress and Her Empire*, loc. 1725, Kindle.

191 **"wonderfully light brooch"**: "An Empire's Relics: A Sovereign's Jewels—Sale of the Trinkets of the Empress of the French, From the *London Daily News*, Dec. 25," reprinted in *New York Times*, January 21, 1872.

191 **"it is the most beautiful"**: François du Buysson, *L'Orchidophile: traité théorique et pratique sur la culture des Orchidées* (1878), 256–57; translation is mine.

194 **"The flowers most rare"**: Charles Baudelaire, part of a stanza of "Invitation to the Voyage," *Les Fleurs du Mal* (*Flowers of Evil*; 1857); translated by Jack Collings Squire, *Poems and Baudelaire Flowers* (London: New Age Press, 1909).

195 **And quite the "picturesque pile"**: George King Matthews, *Abbotsford and Sir Walter Scott* (London: Mabbot, 1854), 30.

196 **"writing-table was encircled"**: Filon, *Recollections of the Empress Eugénie*, 62.

198 **"for taste and workmanship combined"**: "An Empire's Relics: A Sovereign's Jewels—Sale of the Trinkets of the Empress of the French, From the *London Daily News*, Dec. 25," *New York Times*, January 21, 1872, 6.

199 **"The movement of the Queen"**: Ethel Smyth, *Streaks of Life* (New York: Alfred Knopf, 1922), 127.

Ch. 12 The Historian, the Actor, and the Healing of Orchids

201 **One person with this kind:** John Hope Franklin, *Mirror to America: The Autobiography of John Hope Franklin* (New York: Farrar, Straus, and Giroux, 2005), 3, 181, 183.

203 **While John taught at Brooklyn:** John Whittington Franklin, personal interview, October 13, 2021.

203 **The next year, the family:** John Whittington Franklin, personal interview, October 13, 2021.

205 **There, they built their "dream greenhouse":** John Whittington Franklin, personal interview, October 13, 2021.

205 **With begonias and ferns:** Peter Applebome, "Keeping Tabs on Jim Crow; John Hope Franklin," *New York Times Magazine,* April 23, 1995, Section 6, p. 34.

205 **Pigors, who attained professional:** Hermann Pigors, personal communication via email, November 26, 2021. The originators and registrants of named orchids are found in OrchidWiz Encyclopedia database, vers. 7.2 (2021).

205 **It is a small, freckled, bright:** Allen G. Breed, "At 90, Historian Franklin Still Strives to Unite the Two Americas," *Pittsburgh Post-Gazette*, November 5, 2005, https://www.post-gazette.com/ae/books/2005/11/06/At-90-historian-Franklin-still-strives-to-unite-the-2-Americas/stories/200511060175.

206 **Professor and author Cathy Davidson:** Cathy Davidson, personal communication via email, November 4, 2021.

207 **"the history, the provenience":** John Hope Franklin, "Orchidomania" (Book review of Jack Kramer, *Orchids: Flowers of Romance and Mystery*), *The American Scholar* 45, no. 3 (Summer 1976), 460–62.

209 **Psychiatrist Sue Stewart-Smith:** Sue Stewart-Smith, *The Well-Gardened Mind: The Restorative Power of Nature* (Scribner, 2020), 13.

209 **"hybridizing orchids is like wood":** Deb Boersma, "Tiny Slippers, a.k.a. Miniature Paphiopedilums," *Orchids* 90, no. 10 (October 2021), 778.

211 **Although Burr always came across:** Christopher Looby, "Flowers of Manhood: Race, Sex and Floriculture from Thomas Wentworth Higginson to Robert Mapplethorpe," *Criticism* 37, no. 1 (Winter 1995), 145; George Chauncey, *Gay New York: Gender, Urban Culture, and the Making of the Gay Male World, 1890–1940* (Basic Books, 1994), 15.

213 **"the fact is, flowers are":** Judson Hand, "Burr Sticks to His Orchids," *New York Daily News*, July 22, 1977.

213 **In orchids, this trait:** Ernest Hetherington, "Cattleya Hybrids and Hybridizers: Blue Cattleyas," *American Orchid Society Bulletin* 54, no. 5 (May 1985), 550.

215 **Upon the move, Burr donated:** Raymond Burr Orchid Collection in Box 3, Don B. Huntley College of Agriculture Records, Collection no. 0017, University Archives, Special Collections and Archives, University Library, California State Polytechnic University, Pomona.

215 **In 1986, *Orchids* magazine:** Ernest Hetherington, "Cattleya Hybrids and Hybridizers: Prospects for the Future," *Orchids* (May 1986), 452–61.

215 **Jackie Lacey, former president:** John Casey, "The Significance of Flowers in the LGBTQ+ Community," quoting Jackie Lacey, former president of the American Institute of Floral Designers, https://www.1800flowers.com/blog/celebrate-occasions/seasonal-trends/flowers-in-lgbtq-community/. Accessed 10/7/2021.

216 **"Like a suspension in time":** Sue Stewart-Smith, *The Well-Gardened Mind: The Restorative Power of Nature* (Scribner, 2020), 16.

Ch. 13 Orchid Art and Conservation

222 **"had secured about ten thousand":** Albert Millican, *Travels and Adventures of an Orchid Hunter* (London: Castle, 1891), 151.

223 **the desperate need for evocative:** H.R.H. The Duke of Edinburgh, K. G., K. T., foreword to Margaret Mee's *Margaret Mee: In Search of Flowers of the Amazon Forests, Diaries of an English Artist Reveal the Beauty of the Vanishing Rainforest* (Suffolk, England: Nonesuch Expeditions, 1988), 9.

225 **"put aside all other ideas":** Mee, *Margaret Mee* (1988), 24.

229 **"the outstanding find, the rare":** Mee, *Margaret Mee*, 89.

229 *Cattleya violacea* **"were so brilliant":** Mee, *Margaret Mee*, 103.

231 **"The forest was destroyed":** Mee, *Margaret Mee*, 182.

231 **The water was high:** Ruth Stiff, *Margaret Mee: Return to the Amazon* (London: Stationery Office, Royal Botanic Gardens, Kew, 1996), 140. Stiff quotes Mee, *In Search of Flowers of the Amazon.*

232 **"Tomorrow is a sad day":** Stiff, *Margaret Mee*, 261.

233 **"powerful impetus":** Richard Schultes, preface to Margaret Mee's *Margaret Mee: In Search of Flowers of the Amazon Forests, Diaries of an English Artist Reveal the Beauty of the Vanishing Rainforest* (Suffolk, England: Nonesuch Expeditions, 1988), 12.

234 **"The human/nature relationship":** Quoted in *Miroir* magazine, Monde Series, "Madeline von Foerster" (October 2019): 100–107.

234 **"we are most alive when":** Meredith Yayanos, "Into the Woods: The Wisdom of Madeline von Foerster," *Coilhouse Magazine* (January 2009), 56.

234 **"Even as there is always":** Yayanos, "Into the Woods," 59.

235 **Jost is a physicist and:** Lou Jost, personal communication via email, November 26, 2021.

226 **"A good photo or painting":** Lou Jost, personal communication via email, November 26, 2021.

237 **Recent studies on the decline:** Barnabas C. Seyler, Orou G. Gaoue, Ya Tang, and David C. Duffy, "Understanding Knowledge Threatened by Declining Wild Orchid Populations in an Urbanizing China (Sichuan)," *Environmental Conservation* 46 (2019): 318–25; see also Barnabas C. Seyler, Orou G. Gaoue, Ya Tang, David C. Duffy, and Ercong Aba, "Collapse of Orchid Populations Altered Traditional Knowledge and Cultural Valuation in Sichuan, China," *Anthropocene* 29 (2020): 1–12.

Selected Bibliographies by Chapter

❧

Ch. 1 Lusty Ladies of the Enlightenment

Benjamin, Marina, ed. *Science and Sensibility: Gender and Scientific Enquiry 1780–1945*. Cambridge, MA: Basil Blackwell, 1991.

Bewell, Alan. *Natures in Translation: Romanticism and Colonial Natural History*. Baltimore: Johns Hopkins University Press, 2017.

Bewell, Alan. "Erasmus Darwin's Cosmopolitan Nature." *ELH* 76, no. 1 (Spring 2009): 19–48.

Branson, Susan. "Flora and Femininity: Gender and Botany in Early America." *Commonplace: The Journal of Early American Life*. Accessed November 22, 2021, http://commonplace.online/article/flora-femininity/

Browne, Janet. "Botany for Gentlemen: Erasmus Darwin and *The Loves of the Plants*." *Isis* 80, no. 4 (1989): 593–621.

Darwin, Erasmus. *The Botanic Garden, A Poem in Two Parts. Pt. 1 Containing the Economy of Vegetation, Pt. 2 The Loves of the Plants. With Philosophical Notes.* London: J. Johnson, 1791.

Darwin, Erasmus. *Phytologia, or, the Philosophy of Agriculture and Gardening.* London: J. Johnson, 1800.

Darwin, Erasmus. *The Temple of Nature.* London: J. Johnson, 1803.

Dash, Mike. *Tulipomania: The Story of the World's Most Coveted Flower & the Extraordinary Passions It Aroused.* New York: Broadway Books, 2001.

Fryer, Peter. *Mrs. Grundy: Studies in English Prudery.* New York: London House & Maxwell, 1963.

Kelley, Theresa. *Clandestine Marriage: Botany and Romantic Culture.* Baltimore: Johns Hopkins, 2012.

King-Hele, Desmond, ed. *Charles Darwin's "The Life of Erasmus Darwin."* Cambridge, MA: Cambridge University Press, 2003. First published 1879.

King-Hele, Desmond. *Erasmus Darwin.* New York: Scribner, 1963.

Polwhele, Richard. *Unsex'd Females.* London: Cadell and Davies, 1798.

Royal Botanic Gardens, Kew. *Hand-list of Orchids Cultivated in the Royal Gardens.* London: Her Majesty's Stationery Office, 1896.

Stewart, Joyce, ed. *Orchids at Kew.* London: Her Majesty's Stationery Office, 1992.

Sweet, Robert. *The British Flower Garden.* London: W. Simpkin and R. Marshall, 1823–1829.

Uglow, Jenny. *The Lunar Men: Five Friends Whose Curiosity Changed the World.* New York: Farrar, Straus and Giroux, 2002.

Ch. 2 Orchids Fit for a Chinese Empress

Barnhart, Richard M. *Peach Blossom Spring: Gardens and Flowers in Chinese Paintings.* New York: Metropolitan Museum of Art, 1983.

Carl, Katharine. *With the Empress Dowager of China.* New York: The Century Co., 1905.

Chang, Jung. *Empress Dowager Cixi: The Concubine Who Launched Modern China.* New York: Anchor, 2013.

Chieh Tzu Yuan Hua Chuan. *The Mustard Seed Garden Manual of Painting.* Translation edited by Mai-Mai Sze. Princeton, NJ: Princeton University Press, 1992. First published 1679.

Fortune, Robert. *A Journey to the Tea Countries of China.* London: J. Murray, 1852.

Goody, Jack. *The Culture of Flowers*. Cambridge, MA: Cambridge University Press, 1993.

Hansen, Eric. *Orchid Fever: A Horticultural Tale of Love, Lust, and Lunacy*. New York: Vintage, 2000.

Keswick, Maggie. *The Chinese Garden: History, Art, and Architecture*. Cambridge, MA: Harvard University Press, 2003.

Princess Der Ling, First Lady in Waiting to the Empress Dowager. *Two Years in the Forbidden City*. New York: Moffat, Yard, 1911.

Qian Xingjian. *Famous Flowers in China*. Shanghai: Shanghai Press, 2010.

Seagrave, Sterling. *Dragon Lady: The Life and Legend of the Last Empress of China*. New York: Vintage, 1992.

Su Hua Ling Chen. *Ancient Melodies*. New York: Universe Books, 1988. First published 1953.

Wolff, Eric. "Chinese Cymbidium Species." *Orchids* 68, no. 7 (July 1999): 682–93.

Zorn, Johannes. *Auswahl schöner und seltener Gewächse als eine Fortsetzung der Amerikanischen Gewächse*. Nurnberg: im Verlag der Raspeschen Handlung, 1795–1798.

Ch. 3 Orchids in the Tenderloin

A.E.W. "The Orchid Show at New York" and "Scenes from the Orchid Show." *The American Florist* 2, no. 40 (April 1, 1887): 318–20.

Dennett, A. S. *Weird & Wonderful: The Dime Museum in America*. New York: NYU Press, 1997.

Eden Musée Monthly Catalog, January 1892. New York: Rich G. Hollman, 1892.

Loring, John. *Tiffany Jewels*. New York: Harry N. Abrams, 1999.

New York Times. "Among Rare Orchids: A Fine Display Attracts a Crowd to the Eden Musée." February 20, 1890, 8.

New York Times. "Beauty in Flowers: An Orchid Exhibition that Is Attractive and Creditable." February 28, 1889, 8.

New York Times. "The Fifth Orchid Show: No Expense to Be Spared in Making It Successful." February 21, 1891, 5.

New York Times. "The Orchid Show: Sixth Annual Exhibition Opened at the Eden Musée." March 3, 1892, 9.

New York Times. "Orchids in Full Bloom." February 17, 1888, 2.

New York Times. "Wonders of the Orchids: An Exhibition of These Curiosities of Nature's Work." March 2, 1887, 2.

"Orchids of the Future." *The American Florist* 2, no. 41 (April 15, 1887): 349.

Sante, Lucy. *Low Life: Lures and Snares of Old New York.* New York: Farrar, Straus, and Giroux, 1991.

Siebrecht & Wadley Rose Hill Nurseries. *Descriptive Supplemental Catalogue of New, Rare and Beautiful Plants.* Harrisonburg, PA: J. Horace McFarland, 1891.

Taplin, Emily Louise. "New York Notes and Comments." *The American Florist* 3, no. 63 (March 15, 1888): 344–47.

Taplin, Emily Louise. "The New York Orchid Show." *The American Florist* 3, no. 62 (March 1, 1888): 319–21.

Taplin, Emily Louise. "The Orchid Show at New York." *The American Florist* 2, no. 39 (March 15, 1887): 293–94.

Warner, Robert, Benjamin Williams, and Thomas Moore. *The Orchid Album.* London: B. S. Williams, 1885.

Ch. 4 Frida Kahlo's Orchid

Ankori, Gannit. *Imagining Her Selves: Frida Kahlo's Poetics of Identity and Fragmentation.* Westport, CT: Praeger, 2002.

Beckert, C. R. "Canhamiana." *American Orchid Society Bulletin* 15, no. 7 (December 1946): 334–39.

Burrus, Christina. *Frida Kahlo: Painting Her Own Reality.* New York: Abrams, 2008.

Chadwick, A. A., and Arthur E. Chadwick, *The Classic Cattleyas.* Portland, OR: Timber Press, 2006.

Davies, Florence. "Wife of the Master Mural Painter Gleefully Dabbles in Works of Art." *The Detroit News,* February 2, 1933.

Hansen, Eric. *Orchid Fever: A Horticultural Tale of Love, Lust, and Lunacy.* New York: Vintage, 2000.

Herrera, Hayden. *Frida: A Biography of Frida Kahlo.* New York: Perennial, 2002.

Herrera, Hayden. *Frida Kahlo: The Paintings.* New York: HarperCollins, 1991.

Place, Ruth Mosher. "Thrilling as a Jungle Movie: Growing Orchids in Detroit," *Detroit News,* March 26, 1939.

Rosenthal, Mark, ed. *Diego Rivera and Frida Kahlo in Detroit.* New Haven, CT: Yale University Press, 2015.

Tuchman, Phyllis. "Frida Kahlo." *Smithsonian Magazine* (November 2002), https://www.smithsonianmag.com/arts-culture/frida-kahlo-70745811/.

Zavala, Adriana, Mia D'Avanza, and Joanna L. Groarke, editors. *Frida Kahlo's Garden*. New York: DelMonico Books, 2015.

Ch. 5 Rafinesque's Strange Collections

Audubon, John James. *Ornithological Biography*. Edinburgh: Neill, 1833.

Boewe, Charles, ed. *A C. S. Rafinesque Anthology*. Jefferson, NC: McFarland, 2005.

Boewe, Charles, ed. *Profiles of Rafinesque*. Knoxville, TN: University of Tennessee Press, 2003.

Boewe, Charles. "Rafinesque among the Field Naturalists." *Bartonia* 54 (1988): 48–58.

Boewe, Charles. "Rafinesque, Constantine Samuel." In *American National Biography*, edited by Garraty and Carnes. New York: Oxford University Press, 1999.

Durrill, Wayne K. "Becoming Rafinesque: Market Society and Academic Reputation in the Early American Republic" *American Nineteenth Century History* 9, no. 2 (June 2008): 123–40.

Flood, Alison. "John James Audubon and the Natural History of a Hoax." *The Guardian*. May 3, 2016. https://www.theguardian.com/books/booksblog/2016/may/03/john-james-audubon-and-the-natural-history-of-a-hoax.

Index Nominum Genericorum (ING), maintained by the Smithsonian Institution.

Lewis, Andrew J. *A Democracy of Facts: Natural History in the Early Republic*. Philadelphia: University of Pennsylvania Press, 2011.

McKelvey, Susan Delano. *Botanical Exploration of the Trans-Mississippi West, 1790–1850*. Jamaica Plain, MA: Arnold Arboretum of Harvard University, 1991.

Memoirs of the Torrey Botanical Club 16 (1921): 270–71.

Pavord, Anna. *The Naming of Names: The Search for Order in the World of Plants*. New York: Bloomsbury, 2005.

Rafinesque, C.S. *Analyse de la Nature, or Tableau de l'univers et des Corps Organises*. Palerme: Aux depens de l'auteur, 1815.

Rafinesque, C. S. *Flora Telluriana*, Parts 1–4. Philadelphia: H. Probasco, 1836–1838.

Rafinesque, C. S. *Medical Flora of the United States*. Philadelphia: Atkinson and Alexander, 1828.

Rafinesque, C. S. *New Flora and Botany of North America*. Philadelphia: H. Probasco, 1836.

Rafinesque, C. S. "School of Flora: Cypripedium luteum," *The Casket* 5, no. 3 (March 1830), 138.

Stuckley, Ronald L. "Opinions of Rafinesque Expressed by His American Botanical Contemporaries." *Bartonia* 52 (1986): 26–41.

Sweet, Robert. *The British Flower Garden.* London: W. Simpkin and R. Marshall, 1823–1829.

Warren, Leonard. *Constantine Samuel Rafinesque: A Voice in the American Wilderness.* Lexington: University Press of Kentucky, 2004.

World Checklist of Selected Plant Families (WCSP) maintained by the Royal Botanic Gardens, Kew (kew.org).

Ziser, Michael. "Constantine Samuel Rafinesque." In *Early American Nature Writers* edited by Daniel Patterson. Westport, CT: Greenwood Press, 2008.

Ch. 6 The Wind Orchid

Craig, Jack E. "Neofinetia falcata, the Japanese Furan." *American Orchid Society Bulletin* 40, no. 2 (February 1971): 114–19.

Fischer, Jason. "Fuukiran, the Living Antique." *Orchid Digest* (April, May, June 2006): 62–71.

Fischer, Jason. "Introduction to Neofinetia." Live webinar, www.orchidweb.com, February 6, 2021. "Neofinetia Repotting and Training." Live webinar, www .orchidweb.com, June 12, 2021.

Hansen, Eric. *Orchid Fever: A Horticultural Tale of Love, Lust, and Lunacy.* New York: Vintage, 2000.

Marcon, Frederico. *The Knowledge of Nature and the Nature of Knowledge in Early Modern Japan.* Chicago: University of Chicago Press, 2015.

Nishiguchi, Ikuo. "Fukiran History" (English translation booklet). Originally published in *Fukiran: Art of Tradition and Evolution.* Japan: NP, 2014.

Nobuo, Tsuji, and Nicole Coolidge Roumaniere (translator). *History of Art in Japan.* New York: Columbia University Press, 2018.

"Oceoclades falcata." *Gartenflora* vol. 15 (1866), 69.

Ohwi, Jisaburo. *Flora of Japan.* Washington, DC: Smithsonian Institution, 1965.

Screech, Timon. *Japan Extolled and Decried: Carl Peter Thunberg and the Shogun's Realm, 1775–1796.* New York: Routledge, 2005.

Sims, John. *Curtis's Botanical Magazine* vol. 44–46. London: Stephen Couchman, 1817–1819.

Spongberg, Stephen A. *A Reunion of Trees: The Discovery of Exotic Plants and Their Introduction into North American and European Landscapes.* Cambridge, MA: Harvard University Press, 1990.

Suzuki, Kichigoro. "Japanese Orchids—*Neofinetia falcata* and *Ponerorchis gramnifolia*" *American Orchid Society Bulletin* 54, no. 3 (March 1985): 277–81.

Tachibana, Yasukuni. *Ehon Noyamagusa.* Naniwa: Shibukawa Seiemon, 1755.

Thunberg, Carolus Petrus. *Icones Plantarum Japonicum.* Upsaliae: Litteris Johann Frederick Edman, 1794.

Thunberg, Charles Peter. "Botanical Observations on the Flora Japonica" (Read to the Linnaean Society of London October 1, 1793). Reprinted in *Transactions of the Linnaean Society of London* Vol. II (London, 1794).

Thunberg, Charles Peter. *Voyage de C. P. Thunberg au Japon.* 1796.

Ch. 7 The Science of Freedom and Charles Darwin's "Little Book on Orchids"

Alcock, John. *An Enthusiasm for Orchids: Sex and Deception in Plant Evolution.* Oxford: Oxford University Press, 2006.

Allan, Mea. *Darwin and His Flowers: The Key to Natural Selection.* New York: Taplinger, 1977.

Bellon, Richard. "Inspiration in the Harness of Daily Labor: Darwin, Botany, and the Triumph of Evolution, 1859–1868." *Isis* 102, no. 3 (September 2011): 393–420.

Costa, James T. *Darwin's Backyard: How Small Experiments Led to a Big Theory.* New York: W. W. Norton, 2017.

Darwin, Charles. *On the Origin of Species, By Means of Natural Selection, or, The Preservation of Favored Races in the Struggle for Life.* London: John Murray, 1859. Jim Endersby, ed. Cambridge: Cambridge University Press, 2009.

Darwin, Charles. *The Various Contrivances by Which Orchids are Fertilized By Insects.* New York: D. Appleton, 1877. First published 1862.

Darwin, Francis. *The Life and Letters of Charles Darwin, Including an Autobiographical Chapter, Volume 1.* London: John Murray, 1887.

Ellis, William. *Three Visits to Madagascar.* London: John Murray, 1858.

Endersby, Jim. *Orchid: A Cultural History.* Chicago: University of Chicago Press, 2016.

Fuller, Randall. *The Book That Changed America: How Darwin's Theory of Evolution Ignited a Nation.* New York: Viking, 2017.

Lindley, John. *The Vegetable Kingdom: The Structure, Classification, and Uses of Plants*, 3rd edition. London: Bradbury & Evans, 1853.

Litchfield H. E., ed. *Emma Darwin, Wife of Charles Darwin. A Century of Family Letters, Volume 2*. Cambridge: Cambridge University Press, 1904.

Stewart, Joyce, Johan Hermans, and Bob Campbell. *Angraecoid Orchids: Species from the African Region*. Portland, OR: Timber Press, 2006.

Thomson, Keith. "Charles Darwin: The Complete Naturalist." In *The Great Naturalists*. Robert Huxley, ed. London: Thames and Hudson, 2007.

Voskuil, Lynn. "Victorian Orchids and the Forms of Ecological Society." In *Strange Science: Investigating the Limits of Knowledge in the Victorian Age*. Laura Pauline Karpenko and Shalyn Rae Claggett, eds. Ann Arbor, MI: University of Michigan, 2017.

Wallace, Alfred Russel. "Creation by Law." *Quarterly Journal of Science* 4 no. 16 (1867): 471–88.

Ch. 8 Itinerant Orchids, Enslaved People

Abreu-Runkel, Rosa. *Vanilla: A Global History*. London: Reaktion, 2020.

Aguilera, Citali. "Mexican Scientist Saves Vanilla: Exploitation, Genetic Erosion, and Climate Change Have Endangered Vanilla in Mexico." *El Universal*, May 23, 2019. https://www.eluniversal.com.mx/english/mexican-scientist-fighting-save-vanilla.

Alrich, Peggy, and Wesley Higgins. "*Vanilla*: An Expensive Spice." *Orchids* 87, no. 5 (May 2018): 350–53.

Burnett, Joseph. *About Vanilla*. Boston: Joseph Burnett, 1900.

Cameron, Ken. *Vanilla Orchids: Natural History and Cultivation*. Portland, OR: Timber Press, 2011.

Ecott, Tim. *Vanilla: Travels in Search of the Ice Cream Orchid*. New York: Grove Press, 2004.

Havkin-Frankel, Daphna, and Faith C. Belanger, eds. *Handbook of Vanilla Science and Technology*. Hoboken, NJ: Wiley Blackwell, 2018.

Kreziou, Hanna, Hugo de Boer, and Barbara Gravendeel. "Harvesting of Salep Orchids in North-Western Greece Continues to Threaten Natural Populations." *Oryx* 50, no. 3 (July 2016): 393–396.

Rain, Patricia. *Vanilla: The Cultural History of the World's Favorite Flavor and Fragrance*. New York: Tarcher, 2004.

Schmidt, Barbara. "*Vanilla planifolia*: Not Just Another Pretty Flower." *Orchids* 88, no. 3 (March 2019): 204–7.

"Vanilla Beans." Observatory of Economic Complexity, data from 2019. Accessed December 6, 2021, https://oec.world/en/profile/hs92/vanilla-beans #exporters-importers.

Ward, Artemas. *The Grocer's Encyclopedia.* New York: n.p., 1911.

Ch. 9 Jane Loudon and Her Floriferous Press

Chronological list of referenced works by Jane Webb Loudon (only first publication listed):

Jane Wells Webb. *Prose and Verse.* London: R. Wrightson, 1824.

Anonymous [Jane Wells Webb]. *The Mummy! Or a Tale of the Twenty-Second Century.* London: Henry Colburn, 1827.

Anonymous [Jane Loudon]. *The Young Lady's Book of Botany: A Popular Introduction to that Delightful Science.* London: Robert Tyas, 1838.

Mrs. Loudon. *Instructions in Gardening for Ladies.* London: John Murray, 1840.

Mrs. Loudon. *Botany for Ladies, or, a Popular Introduction to the Natural System of Plants.* London: John Murray, 1842.

Mrs. Loudon. *British Wild Flowers.* London: William Smith, 1844.

Mrs. Loudon. *Lady's Country Companion, or, How to Enjoy a Country Life Rationally.* London: Longman, Brown, Green, and Longmans, 1845.

Mrs. Loudon. *Ladies' Companion to the Flower Garden.* London: William Smith, 1846.

Parley, Peter [Jane Loudon]. *Tales about Plants.* London: William Tegg, 1853.

Referenced works by authors other than Jane Webb Loudon:

Bilston, Sarah. "Queens of the Garden: Victorian Women Gardeners and the Rise of the Gardening Advice Text." *Victorian Literature and Culture* 36, no. 1 (2008): 1–19.

Boniface, Priscilla, ed. *In Search of English Gardens: The Travels of John Claudius Loudon and His Wife Jane.* London: Guild Publishing, 1987.

Desmond, Ray. "Victorian Gardening Magazines." *Garden History* 5, no. 3 (Winter 1977): 47–66.

Dewis, Sarah. *The Loudons and the Gardening Press: A Victorian Cultural Industry.* Surrey, England: Ashgate, 2014.

Howe, Bea. *Lady with Green Fingers: The Life of Jane Loudon.* London: Country Life Limited, 1961.

Loudon, John Claudius. *An Encyclopedia of Gardening.* London: Longman Rees Orme Brown Green, 1828.

Paxton's Magazine of Botany and Register of Flowering Plants 5 and 6. London: W. S. Orr, 1839, 1840.

Valen, Dustin. "On the Horticultural Origins of Victorian Glasshouse Culture." *Journal of the Society of Architectural Historians* 75, no. 4 (December 2016): 403–23.

Voskuil, Lynn. "Victorian Orchids and the Forms of Ecological Society." In *Strange Science: Investigating the Limits of Knowledge in the Victorian Age.* Laura Pauline Karpenko and Shalyn Rae Claggett, eds. Ann Arbor: University of Michigan, 2017.

Ch. 10 Orchids and Steel

American Gardening 21, no. 278 (April 21, 1900).

American Gardening 25, no. 475 (1904).

Denker, Ellen Paul. *Faces & Flowers: Painting on Lenox China.* Richmond, Virginia: University of Richmond Museums, 2009.

The Florists' Exchange 46 (1918).

Gardeners' Chronicle (1903, ii).

Gardening 6, no. 126 (Dec. 1, 1897).

"Jay Gould's Orchids." *Placer Herald* 44, no. 45 (September 26, 1896).

Journal of the Horticultural Society of New York. New York: Horticultural Society of New York, 1911, 1912, 1913, 1914.

Kahn, Eve M. "Tiffany Show Reveals Helen Gould's Role as Arts Patron." *New York Times*, June 12, 2018. https://www.nytimes.com/2018/06/12/arts/design/louis-comfort-tiffany-helen-gould-lyndhurst.html?searchResultPosition=1.

Koopowitz, Harold. *Tropical Slipper Orchids: Paphiopedilum and Phragmipedium Species and Hybrids.* Portland, OR: Timber Press, 2008.

McCullough, David. *The Great Bridge: The Epic Story of the Building of the Brooklyn Bridge.* New York: Simon & Schuster, 2001.

New York Times. "All Society in Costume: Mrs. W. K. Vanderbilt's Great Fancy Dress Ball." March 27, 1883, 1.

New York Times. "The Orchid Craze at Its Height in Fashionable New York." February 17, 1907, 2.

New York Times. "Vanderbilt-French Wedding: Floral Decorations to be More Elaborate than First Intended." January 10, 1901, 7.

New York Times. "Vanderbilt Wedding Plans Completed: House to be Decorated Throughout with Orchids in All Colors." January 26, 1908, 9.

The Orchid Review 2 (November 1894), 324–25.

The Orchid Review 11 (October 1903), 311–12.

The Orchid Review 27 (January–February 1919), 33–34.

"Orchids the Favorite." *The Journal* (New York, NY), March 28, 1896, 11.

Schuyler, Hamilton. *The Roeblings: A Century of Engineers, Bridge-Builders and Industrialists.* New York: AMS Press, 1931.

Transactions of the Massachusetts Horticultural Society for the Year 1910. Boston, 1911. 132–38.

Warner, Williams, and Moore. *Orchid Album.* 1893.

Ch. 11 The Flowers, Fashion, and Friendships of Empress Eugénie

Aronson, Theo. *Queen Victoria and the Bonapartes.* London: Thistle Publishing, 2014.

Baudelaire, Charles. *Les Fleurs du Mal (Flowers of Evil)*, 1857. Translated by Jack Collings Squire, *Poems and Baudelaire Flowers.* London: The New Age Press, 1909.

Beaton, D. "The Emperor and Empress of the French and the Queen of England at the Crystal Palace." *Cottage Gardener and Countryman's Companion* 14, no. 344 (May 1, 1855): 65–69.

Beaton, D. "Horticultural Fete at the Crystal Palace." *Cottage Gardener and Countryman's Companion* 14, no. 350 (June 12, 1855): 176–81.

Buckridge, Steeve O. *The Language of Dress: Resistance and Accommodation in Jamaica, 1750–1890.* Kingston, Jamaica: University of the West Indies Press, 2009.

Madame Carette (nee Bouvet). *My Mistress, the Empress Eugenie, or Court Life at the Tuileries.* London: Dean and Son, 1890.

du Buysson, François. *L'Orchidophile: Traité Théorique et Pratique Sur la Culture des Orchidées.* Paris: Auguste Goin, 1878.

de Puydt, Emile. *Les Orchidées, Histoire Iconographique, Organographie, Classification, Géographie, Collections, Commerce, Emploi, Culture, Avec Une Revue Descriptive des Espèces Cultivées en Europe; Ouvrage Orné de 244 Vignettes et de 50 Chromolithographics.* Paris: J. Rothschild, 1880.

Filon, Augustin. *Recollections of the Empress Eugénie.* London: Castle and Company, 1920.

The Scotsman. "Grand Floral Fete at the Crystal Palace." June 6, 1855, 2.

Irving, Washington. *Tales of the Alhambra.* Philadelphia: Lea & Carey, 1832.

Kalba, Laura Anne. "Blue Roses and Yellow Violets: Flowers and the Cultivation of Color in Nineteenth-Century France." *Representations* 120, no. 1 (Fall 2012): 83–114.

Loring, John. *Tiffany Jewels.* New York: Harry N. Abrams, 1999.

Matthews, George King. *Abbotsford and Sir Walter Scott.* London: Mabbot, 1854.

McQueen, Alison. *Empress Eugénie and the Arts: Politics and Visual Culture in the Nineteenth Century.* Surrey, United Kingdom: Ashgate, 2011.

McSweeney, Anna. "Versions and Visions of the Alhambra in the Nineteenth Century Ottoman World." *West 86th* 22, no. 1, 44–69.

New York Times, "An Empire's Relics: A Sovereign's Jewels—Sale of the Trinkets of the Empress of the French, From the *London Daily News,* Dec. 25." January 21, 1872, 6.

Ormond, Richard, and Carol Blackett-Ord, eds. *Franz Xaver Winterhalter and the Courts of Europe, 1830–70.* Harry N. Abrams, 1988.

Seward, Desmond. *Eugénie: The Empress and Her Empire.* London: Thistle Publishing, 2013.

Smyth, Ethel. *Streaks of Life.* New York: Alfred Knopf, 1922.

Ch. 12 The Historian, the Actor, and the Healing of Orchids

Ades, Brian. Sotheby's International Realty. "1830 N. Sierra Bonita, The Raymond Burr Estate." Accessed on 10/12/2021, https://legaciesofla .com/1830-n-sierra-bonita-raymond-burr-estate/.

Breed, Allen G. "At 90, Historian Franklin Still Strives to Unite the Two Americas." *Pittsburgh Post-Gazette,* November 5, 2005. https://www.post-gazette .com/ae/books/2005/11/06/At-90-historian-Franklin-still-strives-to-unite -the-2-Americas/stories/200511060175.

Casey, John. "The Significance of Flowers in the LGBTQ+ Community." Accessed 10/7/2021, https://www.1800flowers.com/blog/celebrate-occasions/ seasonal-trends/flowers-in-lgbtq-community/.

Chadwick, Arthur. "*Cattleya bowringiana*: The Autumn Pixie." November 1, 2015. https://chadwickorchids.com/content/cattleya-bowringiana.

Chadwick, A. A., and Arthur E. Chadwick. *The Classic Cattleyas.* Portland, OR: Timber Press, 2006.

Chauncey, George. *Gay New York: Gender, Urban Culture, and the Making of the Gay Male World, 1890–1940.* New York: Basic Books, 1994.

Endersby, Jim. "Deceived by Orchids: Sex, Science, Fiction, and Darwin." *British Journal for the History of Science* 49, no. 2 (June 2016): 205–29.

Franklin, John Hope. *Mirror to America: The Autobiography of John Hope Franklin.* New York: Farrar, Strauss, and Giroux, 2005.

Franklin, John Hope. "Orchidomania." *The American Scholar* 45, no. 3 (Summer 1976): 460–62.

Franklin, John Whittington. Personal interview by telephone. October 13, 2021.

Hand, Judson. "Burr Sticks to His Orchids." *New York Daily News*, July 22, 1977.

Hetherington, Ernest. "Cattleya Hybrids and Hybridizers: Blue Cattleyas." *American Orchid Society Bulletin* 54, no. 5 (May 1985): 543–52.

Hetherington, Ernest. "Cattleya Hybrids and Hybridizers: Prospects for the Future." *American Orchid Society Bulletin* 55, no. 5 (May 1986): 452–61.

Lombardi, Paula. "Face to Face with Raymond Burr." *Healdsburg Tribune, Enterprise and Scimitar* 116th Year, No. 3 (October 14, 1981).

Looby, Christopher. "Flowers of Manhood: Race, Sex and Floriculture from Thomas Wentworth Higginson to Robert Mapplethorpe." *Criticism* 37, no. 1 (Winter 1995): 109–56.

Janes, Dominic. *Freak to Chic: "Gay" Men In and Out of Fashion after Oscar Wilde.* New York: Bloomsbury, 2021.

Mersmann, Andrew. "Robert Benevides of the Raymond Burr Winery." *Passport Magazine*, Accessed 10/7/2021, https://passportmagazine.com/robert-benevides-of-the-raymond-burr-winery/.

OrchidWiz Encyclopedia database, vers. 7.2 (2021).

Raymond Burr Orchid Collection. Box 3, Don B. Huntley College of Agriculture Records, Collection no. 0017, University Archives, Special Collections and Archives, University Library, California State Polytechnic University, Pomona.

Simpson, Sandra. "Orchid Fancier: Raymond Burr." Tauranga Orchid Society blog, posted April 28, 2020, https://taurangaorchids.wordpress.com/2020/04/28/orchid-fancier-raymond-burr/.

Starr, Michael. *Hiding in Plain Sight: The Secret Life of Raymond Burr.* New York: Applause, 2008.

Stewart-Smith, Sue. *The Well-Gardened Mind: The Restorative Power of Nature*. New York: Scribner, 2020.

Ch. 13 Orchid Art and Conservation

Endersby, Jim. *Orchid: A Cultural History*. Chicago: University of Chicago Press, 2016.

Jost, Lou. "Orchid Conservation: Why? Where? How?" online lecture. Orchid Digest International Speakers Day, October 31, 2020.

"Madeline von Foerster." *Miroir* magazine, Monde Series (October 2019): 100–107.

Mee, Margaret. *Margaret Mee: In Search of Flowers of the Amazon Forest*. Suffolk, England: Nonesuch Expeditions, 1988.

Millican, Albert. *Travels and Adventures of an Orchid Hunter*. London: Castle, 1891.

Porter, Charlotte. "Wetlands and Wildlife: Martin Johnson Heade in Florida." *The Florida Historical Quarterly* 88, no. 3 (Winter 2010): 326–47.

Seyler, Barnabas C., Orou G. Gaoue, Ya Tang, and David C. Duffy. "Understanding Knowledge Threatened by Declining Wild Orchid Populations in an Urbanizing China (Sichuan)." *Environmental Conservation* 46 (2019): 318–25.

Seyler, Barnabas C., Orou G. Gaoue, Ya Tang, David C. Duffy, and Ercong Aba. "Collapse of Orchid Populations Altered Traditional Knowledge and Cultural Valuation in Sichuan, China." *Anthropocene* 29 (2020): 1–12.

Stiff, Ruth. *Margaret Mee: Return to the Amazon*. London: Stationery Office, Royal Botanic Gardens, Kew, 1996.

Yayanos, Meredith. "Into the Woods: The Wisdom of Madeline von Foerster." *Coilhouse Magazine* (January 2009): 56.

Illustration Credits

86   Photo by the author

89   Photo by the author

90   Courtesy Smithsonian Libraries

91   © The Trustees of the British Museum. All Rights Reserved.

92   F.H. King, *Farmers of Forty Centuries, or, Permanent Agriculture in China, Korea, and Japan* (1911)

93   Courtesy HathiTrust

94   The Picture Art Collection / Alamy Stock Photo

96   Courtesy HathiTrust

97   Courtesy Biodiversity Heritage Library

98   Courtesy HathiTrust

100 Album / Alamy Stock Photo

103 Photo by the author

106 Courtesy Wikimedia Commons. From Francis Darwin, *The Life and Letters of Charles Darwin* (1887)

107 Gareth Christian

108 Courtesy Wikimedia Commons

109 Gareth Christian

111 Courtesy Biodiversity Heritage Library

112 Courtesy Biodiversity Heritage Library

113 Courtesy Biodiversity Heritage Library

116 Courtesy Biodiversity Heritage Library

120 Coulanges/Shutterstock.com

122 Rui Santos/Shutterstock.com

128 © Vanillinmacher/WikiMedia CC attribution only license

129 Arterra Picture Library / Alamy Stock Photo

130 Courtesy Library of Congress

131 Courtesy Schomburg Center for Research in Black Culture, Photographs and Prints Division, The New York Public Library

132 Courtesy Internet Archive (internetarchive.org)

134 Prachaya Roekdeethaweesab/Shutterstock.com

135 Author's collection

136 Courtesy Biodiversity Heritage Library

140 © Jason Terry Studio

143 Antiquarian Images / Alamy Stock Photo

144 The History Collection / Alamy Stock Photo

145 Courtesy Biodiversity Heritage Library

147 Courtesy Biodiversity Heritage Library

150 Courtesy Biodiversity Heritage Library

153 Courtesy Biodiversity Heritage Library

155 Courtesy Biodiversity Heritage Library

156 Courtesy Biodiversity Heritage Library

158 Courtesy Biodiversity Heritage Library

159 Paul Atkinson/Shutterstock.com

162 Byron Company. Museum of the City of New York. 93.1.1.9995

163 Bain News Service, publisher. Courtesy Wikimedia Commons

165 William E. Sackett, *Scannell's New Jersey First Citizens: Biographies and Portraits* (Paterson, NJ: J.J. Scannell, 1917–1918)

166 Courtesy Smithsonian American Art Museum, Gift of Mrs. Evan M. Wilson

168 Courtesy Biodiversity Heritage Library

170 Courtesy Biodiversity Heritage Library

171 Courtesy Biodiversity Heritage Library

172 Emmi Mattes

174 "Twelve Lenox Porcelain Gold-Ground Monogrammed Botanical Place Plates, c.1906 (porcelain)." Private Collection. Photo © Christie's Images/Bridgeman Images

175 Nolehace Photography/Bob Lewis

177 Courtesy Smithsonian Institution

182 agefotostock / Alamy Stock Photo

183 Marques / Shutterstock.com

185 Artefact / Alamy Stock Photo

187 Author's collection

188 Chateau de Compiegne, Oise, France/Bridgeman Images

190 "A superb antique diamond brooch on a black background." Private Collection. Photo © Christie's Images/Bridgeman Images

193 Wikimedia Commons

195 Photo by the author

196 Photo by the author

197 Heritage Image Partnership Ltd / Alamy Stock Photo

199 Courtesy Smithsonian Institution

202 STM-035851372, Kevin Horan/Chicago Daily News. © Sun-Times Media, LLC, and Chicago History Museum

204 Courtesy Biodiversity Heritage Library

206 From the collection of John W. Franklin

208 © Estate Hans Erni, Lucerne. Painting resides at Raymond Burr Vineyards, Healdsburg, California

210 © American Orchid Society

214 Courtesy Biodiversity Heritage Library

216 Ron Galella/Ron Galella Collection via Getty Images

217 © Simmie Knox. Photo credit: John Hope Franklin papers, David M. Rubenstein Rare Book & Manuscript Library, Duke University

219 Emmi Mattes

223 Photograph © 2022 Museum of Fine Arts, Boston. Gift of Maxim Karolik for the M. and M. Karolik Collection of American Paintings, 1815–1865. 47.1175

224 Tony Morrison / South American Pictures

226 Courtesy Biodiversity Heritage Library

227 Courtesy Biodiversity Heritage Library

228 © Estate of Margaret Mee and the Board of Trustees of the Royal Botanic Gardens, Kew

230 © Estate of Margaret Mee and the Board of Trustees of the Royal Botanic Gardens, Kew

232 Madeline von Foerster

235 Lou Jost

237 Courtesy Biodiversity Heritage Library

244 © Jason Terry Studio

245 © Jason Terry Studio

246 © Jason Terry Studio

Index

Page numbers in *italics* refer to illustrations.